HP Prime Guide

FINITE MATHEMATICS

For Management, Natural, and Social Sciences

By

LARRY SCHROEDER, MS Mathematics

Formerly Professor of Mathematics, Carl Sandburg College
Author Cengage Learning® and Computer Learning Service

COMPUTER   LEARNING   SERVICES

HP Prime Guide FINITE MATHEMATICS

HP Prime Guide FINITE MATHEMATICS
By LARRY SCHROEDER, MS Mathematics
Formerly Professor of Mathematics, Carl Sandburg College
Author Cengage Learning® and Computer Learning Service

Published by
Computer Learning Service
1013 Woodbine Circle
Galesburg, IL 61401-2358
ComputerLearningService.com

Copyright © 2022 Larry Schroeder

All rights reserved. No part of the HP Prime content in this publication may be reproduced, distributed, or transmitted in any form or by any means, including photocopying, recording, or other electronic or mechanical methods without the prior written permission of the publisher, except in the case of brief quotations in reviews and certain other non-commercial uses permitted by copyright law.

Version 1.0.1

ISBN: 978-0-915573-03-5

The major resource for *HP Prime Guide FINITE MATHEMATICS* is our collection of 90 YouTube videos for finite mathematics. They give specific instructions on implementing the virtual HP Prime or HP Prime calculator for all the screens shown in our book *HP Prime Guide FINITE MATHEMATICS* or OneNote notebook Finite Math. (Handwritten screens in videos have been revised with text screens in this book.)

The free virtual HP Prime calculator material is organized for Finite Math on the Finite Math tab at the Prime Academy | Learning Center website. Use the fly-out menu at the Finite Math tab to go to a HP Prime discussion of specific Finite Math topics.

In the *HP Prime Guide FINITE MATHEMATICS,* we parallel the sections presented in our OneNote's *Finite_Math* notebook. All web-based links shown in this publication are active in our *Finite_Math* notebook. A link to our *Finite_Math* notebook can be found on the home page of our *Prime Academy | Learning Center*.

Dedication

To my wife and children for their love and support.

CONTENTS

Preface   ix

Resources   x

Getting Started   xii

1 CHAPTER Lines   1

Formulas
Point Slope   2
Standard Form   3
Two Point   4
Parallel Perpendicular   5

Application
Appreciation   7
Depreciation   8
Break Even   9
Supply and Demand   10

Method of Least Squares
Equation Best Fit   12
Immediate Mode   13
Future Predications Estimates   14

2 CHAPTER Finance   15

Formulas
Simple Interest   16
Compound Interest   17
Loan   18
Ira   19
Annuity Due   20

Explorations
Compound Interest   22
Continuous Compounding of Interest   23
Effective Rate   25
Sinking Fund   26

3 CHAPTER System of Linear Equations 27

Introductions – System of Linear Equations
One Point 28
Infinitely Many Solutions 29
No Solution 30

Matrix Approach
One Point 32
First Element is a Zero 33
Infinitely Many Solutions 34
No Solution 35

Additional Investigations
Tools 37
Higher Equations Higher Unknowns 39
Fewer Equations Than Unknowns 41
More Equations Than Unknowns 43

Matrices
Matrix Arithmetic 46
Matrix Algebra 48
Multiplication 50
Inverse 51
    Pivmat – Row Operations 52
    Matrix Equation 53
    Not Sufficient 54

4 CHAPTER Linear Programing 55

Graphical Maximums
Bounded – Standard 56
Bounded – Nonstandard 58

Graphical Minimums
Unbounded – Standard 61
    Intercepts 62
    Simultaneous Equations 63
Line – Nonstandard 64

Simplex Maximums
Standard 67
Nonstandard 69

Simplex Minimums
The Dual 72
Line – Nonstandard 74

5 CHAPTER Sets and Probabilities 76

Sets
Set Operations 77
Orientations 78
Set Algebra 80
De Morgan's Law 81

Number Elements Finite Set
Count for A and B 83
Applications 84

Multiplication Principle
Coin Tosses 87
Applications 88

Permutations and Combinations
Introduction 90
Permutations 91
Combinations 92
Explorations 93

Experiment or Observation
Terminology 95
Venn Diagrams 97

Probability
Theoretical and Subjective 99
Additional Topics 100

Rules of Probability
Basic Properties
Computations 101

6 CHAPTER Additional Topics Probabilities   104

Counting Techniques
Additional Investigations   105
Higher Sample Points   106

Conditional Probability
Formulas   108
Independent and Dependent Events   109

Bayes' Theorem
Formula   112

Probability Distributions
Random Variables   114
Histogram   115
7 or 11   116

Expectations
Expected Value   118
Odds   119

Variance and Standard Deviation
Manual   121
App   122
Chebychev's Inequality   123

Binomial Distributions
Bernoulli Trails   125
Formula   126

Normal Distribution Basics
Density Curve   129
Probabilities   130
Z Scores   131

More Normal Distributions
Other Means | Standard Deviations   133
Approximate Binomial Distributions   135

Terms of Use

By using *HP Prime Guide FINITE MATHEMATICS* in any manner, you agree to all of the terms and conditions contained herein.

Trademarks: Cengage Learning ®, Thomson Learning™, WebTutor™ Brooks/Cole are registered trademarks of Cengage Learning Inc. HP Inc. has rights in the registered trademark HP, as well as its other registered and unregistered trademarks. T3™, Teachers Teaching with Technology™, TI-Nspire™, and TI-Navigator™ are registered trademarks of Texas Instruments Inc. iPhone ®, iPad ®, iPod ®, and Mac ® are registered trademarks of Apple Inc. All other trademarks are the property of their respective owners. Any usage of these terms anywhere throughout this book is done so simply as part of a description of the product. This book is in no way affiliated with or endorsed by Cengage Learning, HP, Texas Instruments, Apple or any other product or vendor mentioned in this book.

LIMIT OF LIABILITY/DISCLAIMER OF WARRANTY: THE PUBLISHER AND THE AUTHOR MAKE NO REPRESENTATIONS OR WATTANTIES WITH RESPECT TO THE ACCURACY OR COMPLETENESS OF THE CONTENTS OF THIS WORK AND SPECIFICALLY DISCLAIM ALL WARRANTIES, INCLUDING WITHOUT LIMITATION WARRANTIES OF FITNESS FOR A PARTICULAR PURPOSE. NO WARRANTIES MAY BE CREATED OR EXTENDED BT SALES MATERIALS. THE ADVICE AND STRATEGIES CONTAINED HEREN MAY NOT BE SUITABLE FOR EVERY SITUATION. THIS WORK IS SOLD WITH THE UNDERSTANDING THAT THE PUBLISHER IS NOT ENGAGED IN RENDERING LEGAL, ACCOUNTING, OR OTHER PROFESSIONAL SERVICES. IF PROFESSIONAL ASSISTANCED IS REQUIRED, THE SERVICES OF A COMPETENT PROFESSIONAL PERSON SHOULD BE SOUGHT. NEITHER THE PUBLISHER NOR THE AUTHOR SHALL BE LIABLE FOR DAMAGES, ARISING HEREFROM, THE FACT THAT AN ORGANIZATION OR WEBSITE IS REFERRED TO IN THIS WORK AS A A CITATION AND/OR A POTIENTIAL SOURCE OF FURTHER INFORMATION DOES NOT MEAN THAT THE AUTHOR OR THE PUBLISHER ENDORSES THE INOFRMATION THE ORGANIZATION OR WEBSITE MAY PROVIDE OR RECOMMENDATIONS IT MAY MAKE FURTHER, READER SHOULD BE AWARE THAT INTERNET WEBSITES LISTED IN THIS WORK MY HAVE CHANGED OR DISAPPEARED BETWEEN WHEN THIS WORK WAS WRITTEN AND WHEN IT IS READ.

PREFACE

The *HP Prime Guide Finite Mathematics For Management, Natural, and Social Sciences* is a collection of formulas, definitions, terminology, laws, theorems, rules, properties, principles, graphs, diagrams, charts, distributions, applications, techniques, explorations, investigations, expectations, experiments, trials, and observations that are all shown with HP Prime calculator screens.

The finite mathematics course is littered with burdensome calculations that conceal the underlying math processes and can lead to a premature failure by students in the course. That is where the free virtual HP Prime calculator comes in. It is readily available as a free app to Apple and Android phone users.

The free version works on iPads, tablets, and Chromebooks as well. There is a free emulator for PC and mac owners of the physical calculator. The virtual HP Prime calculator or physical HP Prime calculator turns the course into an understandable and pleasurable experience. When you arrive at the correct solution undeterred by error prone tedious fraction and positive/negative calculations, this can be that pleasurable experience.

We just finished with the perfect place for you to get started with learning your virtual or HP Prime calculator, *HP Prime Guide THE SILVER-BURDETT ARITHMETICS (Annotated) Selected Exercises*. Click Read more at my video's Pinned Comment and follow the links for more information.

I wish to thank *The Silver-Burdett Arithmetics* two authors George Morris Philips and Robert F Anderson, now deceased, for producing their early twentieth century work. We annotated their work, and brought it into the modern era, using the [Home] part of the two-part HP Prime calculator Operating System. The material found in our annotated work is the best place for you to get started.

For the [CAS] part of the HP Prime calculator Operating System we recommend our eBook edition of *HP Prime Guide Algebra Fundamentals*, it is designed to be use on your phone, tablet, computer, or eBook reader. In the *HP Prime Guide Finite Mathematics* [CAS] is not used as much as in Algebra and Calculus courses. The phone version of *HP Prime Guide Algebra Fundamentals* does make a handy reference.

Resources

BASIC MATH ONENOTE NOTEBOOK SUMMARY
    [Arithmetic Fundamentals](#) Prime Academy | Learning Center Website (additional information and instructions for all of OneNote's HP Prime's calculator screens)
    [Playlist of Basic Math Notebook](#) (Learn the calculator features of the HP Prime or virtual HP Prime)
    [General Overview](#) video

GENERAL
HP Prime Guide THE SILVER-BURDETT ARITHMETICS (Annotated) Selected Exercises
    [Amazon](#) - printed Paperback
    [IngramSpark](#) - printed Paperback

HP Prime Guide Algebra Fundamentals
    [Amazon](#) - printed Paperback - Kindle eBook
    [Apple](#) - iBook
    [Barnes and Noble](#) - printed Paperback - Nook
    [Rakuten Kobo](#) - eBook
    [IngramSpark](#) - printed Paperback

YouTube - videos - made with OneNote and virtual HP Prime
    Finite Math – [90 Videos](#)
    Use [Prime Academy | Learning Center's](#) Finite Math drop down flyout to view link to content
    [Example](#) - Finite Math>System of Equations and Matrices>An Introduction to Linear System

CLS HP Prime Academy
    [Prime Academy | Learning Center](#) - website
    Generalities (settings and examples) [HP Prime calculator](#)
    Academy | HP Prime [Table of Contents using Video links](#) (fast internet connection advised)

YouTube
    [HP PRIME 01 - Introduction and Set up](#)
    [Change HP Prime Screen Time](#)

Forum
    [HP Prime Code Snippets and Relative Help](#) for Carbajo's Xcas YouTube Videos

Getting Started

The HP Prime Calculator is two calculators in one. First calculator is the [Home] View calculator. The [Home] View calculator is used in various parts of the tutorial. A second CAS calculator is available. The Computer Algebra System View [CAS] calculator is used for exact calculations. Exact calculations are used extensively in Algebra thru Post Calculus courses. We will have uses for [CAS] in this tutorial as well. CAS appears in the blue title screen when it is being used. For [Home] view this area is blank.

Another major feature of the HP Prime Calculator is its Application Library. At various places in the tutorial, we will use the Function, Advanced Graphing, Statistic 1 Var, and Statistic 2 Var applications.

This tutorial is a collection of basic math examples with explanations covered in a Finite Mathematics course. A review of these examples and accompanying explanations will allow you to have success with a Finite Mathematics course and any future job using these skills.

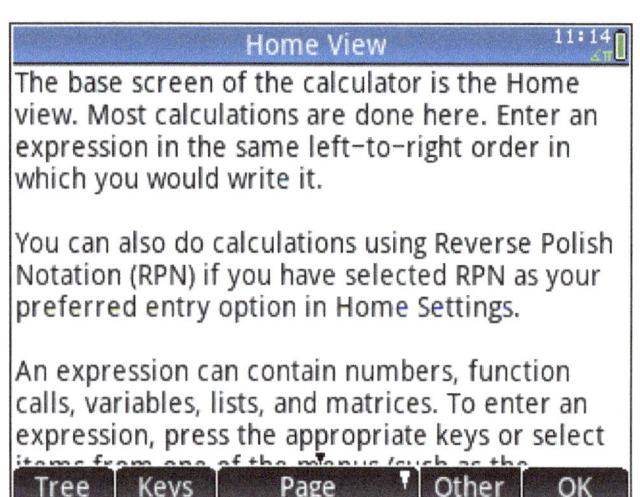

[Home] [Help]. Home View's help is displayed. Notice White shaded down arrow shown on soft Page key.

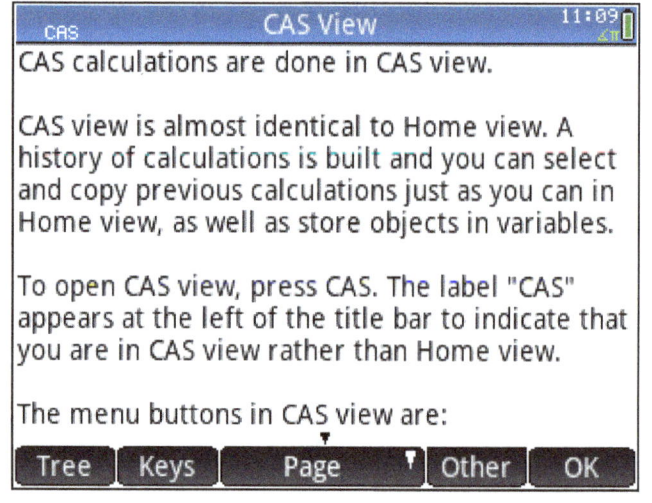

[CAS] [Help]. CAS View's help is displayed. [Esc] or soft key [OK] returns us to the current command line.

The Prime's built-in [Help] feature is a powerful tool. Turn the Prime on. Press [Home] [Help] keys and then soft key [OK] to close. Follow by the [CAS] [Help] keys. From time to time, we will use a HP Prime command that comes from the [Toolbox] key, soft key [Catlg] or soft key [CAS]. Use the Prime's [Help] system to display the command's features and sample examples.

HP Prime Guide

FINITE MATHEMATICS

1

Lines

> Mathematician need only peace of mind and occasionally, paper and pencil."
> ~ Paul Hoffman

Formulas

- Point Slope 2
- Standard Form 3
- Two Point 4
- Parallel Perpendicular 5

Application

- Appreciation 7
- Depreciation 8
- Break Even 9
- Supply and Demand 10

Method of Least Squares

- Equation Best Fit   12
    - ✓ Immediate Mode   13
- Future Predications Estimates   14

Formulas

Sunday, April 21, 2019   3:40 PM

http://computerlearningservice.com/Academy/Finite-Math/Lines/Point---Slope/point---slope.html

Point Slope

POINT SLOPE

$$y - y_1 = m(x - x_1)$$

FUNCTION

$$y = m(x - h) + k$$

Standard Form

STANDARD FORM

1) $x = c$ or   2) $y = mx + b$

GENERAL FORM

$ax + by + c = 0$

slope $= \dfrac{-a}{b}$ y intercept $= \dfrac{-c}{b}$

Two Point

Sunday, April 21, 2019   3:56 PM

$$m = \frac{y_2 - y_1}{x_2 - x_1} \quad y - y_1 = m(x - x_1)$$

USE LETTERS INSTEAD OF SUBSCRIPTS

$$y = \frac{d - b}{c - a}(x - a) + b$$

$$P_1(-4, 3) \quad P_2(5, -2)$$
$$P_1(a, b) \quad P_2(c, d)$$

Parallel Perpendicular

Sunday, April 21, 2019   5:17 PM

$$\text{Parallel} \qquad \text{Perpendicular}$$

$$m_1 = m_2 \qquad m_1 = \frac{1}{m_2}$$

perpendicular to $\quad y = \frac{1}{2}x + 4 \quad P(1,1)$

HP Prime calculator

$$linemb(y, m, x, b) := solve(y = m * x + b, z)$$

$$P(1,1) \quad m = -2$$
$$1 = -2 \cdot 1 + b$$
$$y = -2x + 3$$

$$\frac{1}{2}x + 4 = -2x + 3$$

Steps for Solving Linear Equation

- Multiply −2 and 1 to get −2.
  $1 = -2 + b$

- Swap sides so that all variable terms are on the left hand side.
  $-2 + b = 1$

- Add 2 to both sides.
  $b = 1 + 2$

- Add 1 and 2 to get 3.
  $b = 3$

Applications

Sunday, April 21, 2019   3:50 PM

http://computerlearningservice.com/Academy/Finite-Math/Lines/Applications/applications.html

Appreciation

Sunday, April 21, 2019   5:18 PM

Jim buys an classic car for $50,000. After 20 years it has appreciated to a value of $70,000. Assuming the car appreciates the same each year, write a linear equation giving the value of the car, y, in terms of the number of years Jim has owned it, x.

$$(0, 50), (20, 70)$$
$$y = x + 50$$

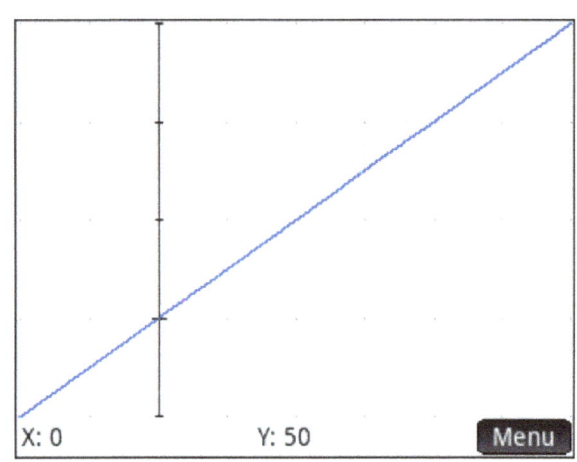
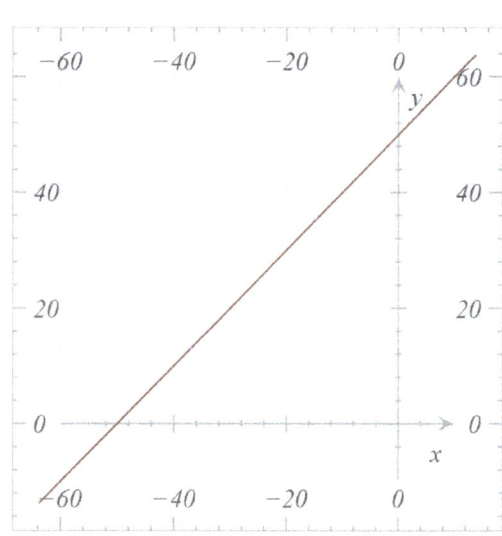

Depreciation

Sunday, April 21, 2019   5:20 PM

A planar in a factory has an original value of $200,000 and is to be depreciated over 5 years with a $50,000 scrap value. Find an equation giving the book value at the end of year x. What will the book value of the planar be at the end of the third year? What is the rate of depreciation of the planar?

$$(0, 200), (5, 50)$$
$$y = -30x + 200$$
$$110,000 \quad -30,000$$

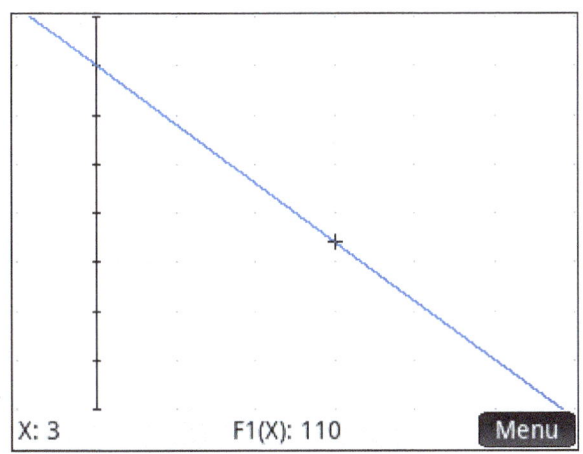
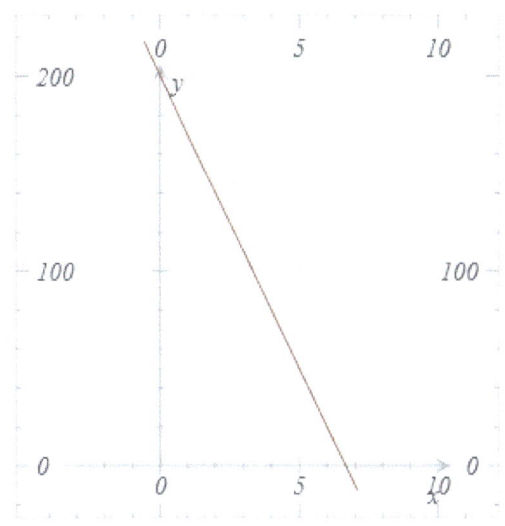

Break Even

Sunday, April 21, 2019   5:21 PM

A company makes an eBook for $3 each and sells the same eBook for $5 each. If the fixed cost of operating the company is $10000, find the break even quantity.

$$C(x) = VC \cdot x + FC \quad R(x) = p \cdot x$$
$$P(x) = R(x) - C(x)$$
$$VC = 3 \quad p = 5 \quad FC = 10{,}000$$

$$y = 3x + 10000 \quad y = 5x \quad \begin{array}{l} y = 5x - (3x + 10000) \\ y = 2x - 10000 \end{array}$$

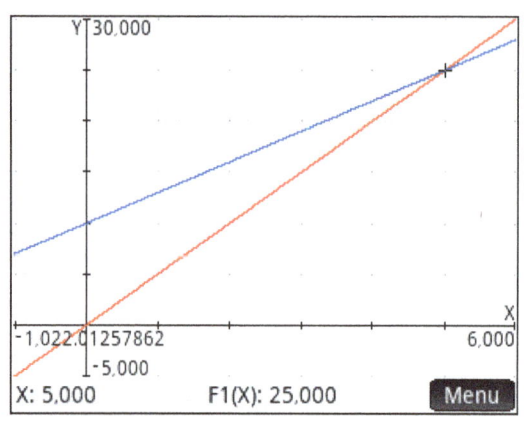

$$3x + 10000 = 5x$$
$$(5000, 25000)$$

Supply and Demand

Sunday, April 21, 2019   5:23 PM

A self-author/publisher is able to sell 400 copies of Genealogy Tips pamphlet a month if the price is $3.50 per pamphlet but sells drop to 300 per month if the price is raised to $4 per pamphlet. At a price of $2 his supply of pamphlets was 150 where at a price of $5 his supply was at 375. Find the demand and supply equations assuming they are linear. Determine the equilibrium price and equilibrium demand/supply.

$$Q_D = (3.50, 400), (4.00, 300)$$
$$Q_S = (2.00, 1.50), (5.00, 375)$$
$$Q_D = -200p + 1100 \quad Q_S = 75p$$

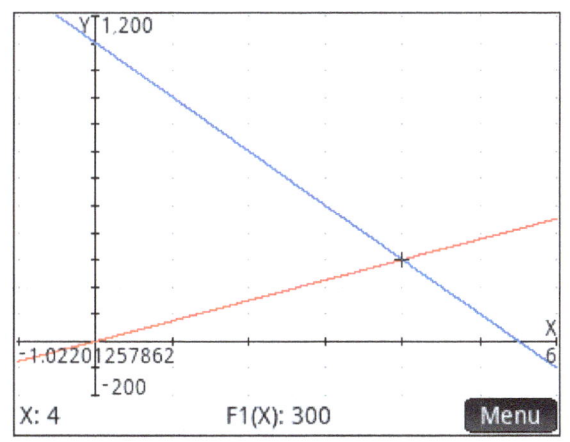

$$P = \$4$$
$$Q_S = Q_D = 300$$
$$-200x + 1100 = 75x$$

Method of Least Squares

Saturday, August 24, 2019   10:45 AM

http://computerlearningservice.com/Academy/Finite-Math/Lines/Method-Least-Sq/method-least-sq.html

Equation Best Fit

Saturday, August 24, 2019   5:56 AM

Method of Least Squares
Given set of data points for *x* and *y*. The least square regression line, $y = f(x) = mx + b$, is where *m* and *b* satisfy the equations

$$\Sigma x \cdot m + n \cdot b = \Sigma y$$
$$\Sigma x^2 \cdot m + \Sigma x \cdot b = \Sigma(x \cdot y)$$

Find the least square line given
$P_1(1,2) \quad P_2(2,3) \quad P_3(3,3) \quad P_4(4,3) \quad P_5(5,5)$

$$15 \cdot m + 5 \cdot b = 16$$
$$55 \cdot m + 15 \cdot b = 54$$

$$(m, b) = (0.6, 1.4)$$

$$y = f(x) = 0.6x + 1.4 = \frac{3}{5}x + \frac{7}{5}$$

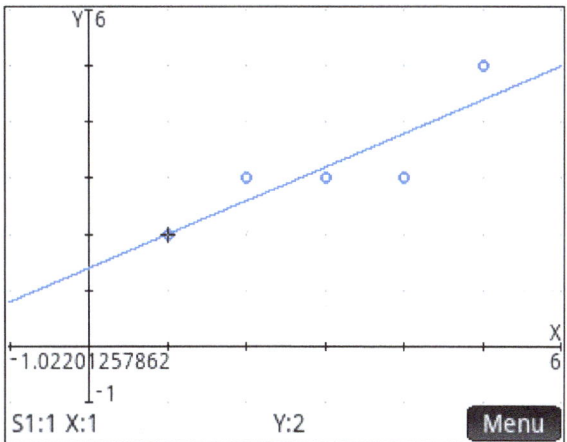

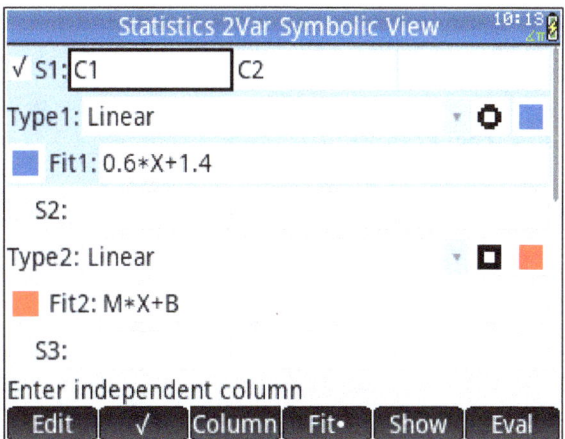

Lines Page 12

Immediate Mode

Sunday, August 25, 2019   5:44 AM

linemb1(C1,C2):
=solve({(sum(C1)*m+SIZE(C1)*b) = (sum(C2)),(sum(C1*C1)*m+sum(C1)*b) = (sum(C1*C2))},{m,b})

linemb2(L1):=L1(1,1)*'x'+L1(1,2)

linemb3(C1,C2):
=RREF([[sum(C1),SIZE(C1),sum(C2)],[sum(C1*C1),sum(C1),sum(C1*C2)]])

linemb4(M1):=M1(1,3)*'x'+M1(2,3)

linear_regression(C1,C2)

linemb5(M1):=M1(1)*'x'+M1(2)

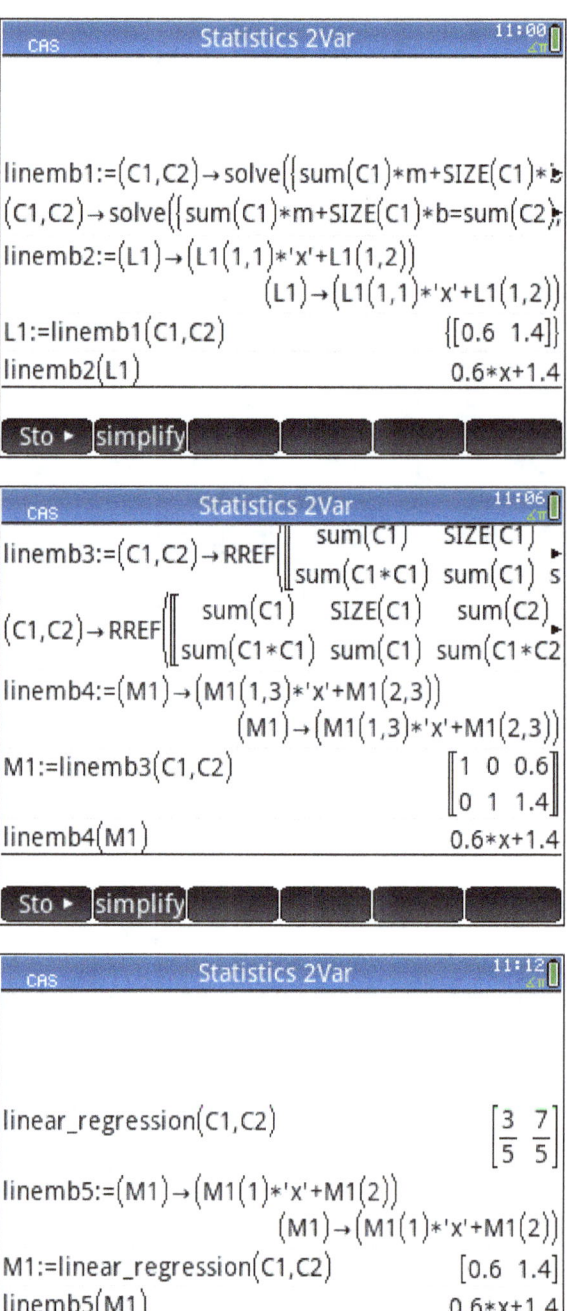

Lines Page 13

Future Predications Estimates

Saturday, August 24, 2019   10:07 PM

A local fleet of Chicago ice scream vendors found how many drumsticks were sold vs how many sailboats from three local marinas were out on Lake Michigan. If 120 sailboats were out, how many drumsticks did they expect to sale? Answer 185

Sailboats	20	30	50	70	90
Drumsticks	40	50	70	100	150

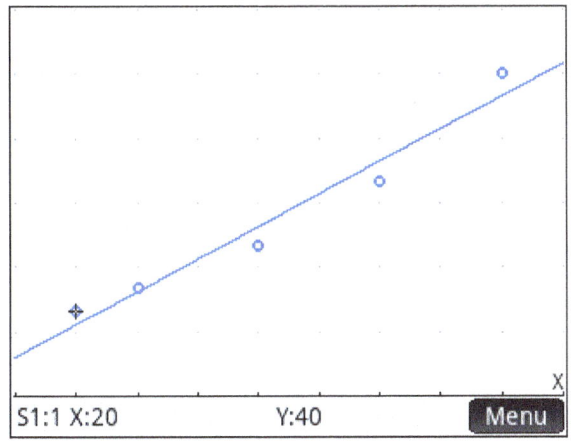

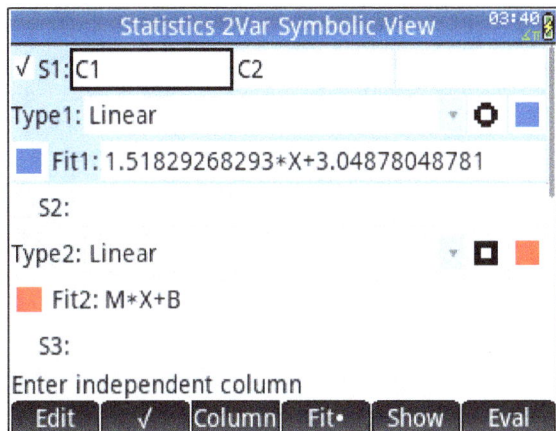

$\Sigma x \cdot m + n \cdot b = \Sigma y$

$\Sigma x^2 \cdot m + \Sigma x \cdot b = \Sigma(x \cdot y)$

$f(120) \approx 185$

$r = \dfrac{n\Sigma(x \cdot y) - \Sigma x \cdot \Sigma y}{\sqrt{[n\Sigma x^2 - (\Sigma x)^2] \cdot \left[n\Sigma y^2 - (\Sigma y)^2\right]}} \approx 0.98$

$260 \cdot m + 5 \cdot b = 410$

$16800 \cdot m + 260 \cdot b = 26300$

$(m, b) = \left(\dfrac{249}{164}, \dfrac{125}{41}\right)$

$y = f(x) \approx 1.52x + 3.05$

Lines Page 14

2

Finance

"If you formulate your question properly, mathematics gives you the answer."
~ Savas Dimopoulos

Formulas

- Simple Interest   16
- Compound Interest   17
- Loan   18
- Ira   19
- Annuity Due   20

Explorations

- Compound Interest   22
- Continuous Compounding of Interest   23
- Effective Rate   25
- Sinking Fund   26

Formulas

Sunday, April 21, 2019   3:42 PM

http://computerlearningservice.com/Academy/Finite-Math/Finance/Formulas/formulas.html

Simple Interest

Sunday, April 21, 2019   9:56 PM

$$I = Prt$$
$$A = P + I = P + Prt$$
$$A = P(1 + rt)$$

$$= 2000\,(1 + 0.05 \cdot 10)$$

$$5000 = P\,(1 + 0.05 \cdot 3)$$

Steps for Solving Linear Equation

- Multiply 0.05 and 10 to get 0.5.
  $A = 2000(1 + 0.5)$

- Add 1 and 0.5 to get 1.5.
  $A = 2000 \cdot 1.5$

- Multiply 2000 and 1.5 to get 3000.
  $A = 3000$

Steps for Solving Linear Equation

- Multiply 0.05 and 3 to get 0.15.
  $5000 = P(1 + 0.15)$

- Add 1 and 0.15 to get 1.15.
  $5000 = P \cdot 1.15$

- Swap sides so that all variable terms are on the left hand side.
  $P \cdot 1.15 = 5000$

- Divide both sides by 1.15.
  $P = \dfrac{5000}{1.15}$

- Expand $\dfrac{5000}{1.15}$ by multiplying both numerator and the denominator by 100.
  $P = \dfrac{500000}{115}$

- Reduce the fraction $\dfrac{500000}{115}$ to lowest terms by extracting and canceling out 5.
  $P = \dfrac{100000}{23}$

Compound Interest

$$FV = PV \cdot \left(1 + \frac{r}{m}\right)^{m \cdot t}$$

$$A = 1000 \cdot \left(1 + \frac{0.05}{12}\right)^{12 \cdot 3}$$

A

$$3000 = P \cdot \left(1 + \frac{0.05}{12}\right)^{12 \cdot 3}$$

$$P = \frac{146125481875413598246519797461850389110697836412928000000000000000000000000000000000000}{565735631628383040361544425867290607739985882206675297054580264308107208365422852566641}$$

Loan

Wednesday, April 24, 2019   1:47 AM

Present Value of an Annuity - Loan

PV of an Annuity, *P*, is the present dollar amount for a stream of equal payments, *R*, of its future current dollar value.

$$PV = PMT \left[\frac{1 - \left(1 + \frac{r}{m}\right)^{-m \cdot t}}{\frac{r}{m}} \right]$$

$$P = 400 \left[\frac{1 - \left(1 + \frac{0.05}{12}\right)^{-12 \cdot 5}}{\frac{0.05}{12}} \right] \qquad 30000 = R \left[\frac{1 - \left(1 + \frac{0.05}{12}\right)^{-12 \cdot 5}}{\frac{0.05}{12}} \right]$$

$$P = 21{,}196.28 \qquad R = 500$$

$$P = 400 \cdot \left(\left(1 - (241/240)^{-60}\right) \cdot 240 \right)$$

$$= \frac{17671859668904550066134233691438625143715436124581308013969890509080189756370484305}{833724481840178623836933255987456062094496206528160480425403079472495588919428331}$$

IRA

Thursday, April 25, 2019   12:54 AM

Future Value of an Annuity - IRA

FV of an Annuity, *A*, is what the dollar amount is at a future date for a stream of equal payments, *R*.

$$FV = PMT\left[\frac{\left(1+\frac{r}{m}\right)^{m\cdot t} - 1}{\frac{r}{m}}\right]$$

$$A = 500\left[\frac{\left(1+\frac{0.0625}{12}\right)^{12\cdot 37} - 1}{\frac{0.065}{12}}\right] = 867{,}764.42$$

$$83060.94 = R\left[\frac{\left(1+\frac{0.0625}{12}\right)^{12\cdot 10} - 1}{\frac{0.065}{12}}\right] \quad R = 500$$

Annuity Due

Tuesday, May 7, 2019   1:21 AM

Future Value of an Annuity - Annuity Due

FV of an Annuity, *A*, is what the dollar amount is at a future date for a stream of equal payments, *R*. Payments made at the beginning of a period.

irabegin(fv,pmt,r,m,t):=ROUND(solve(fv=pmt*(1+r/m)*(((1+r/m)^(m*t)-1)/(r/m)),x),2)

$$FV = PMT \cdot \left(1 + \frac{r}{m}\right) \left[\frac{\left(1 + \frac{r}{m}\right)^{m \cdot t} - 1}{\frac{r}{m}}\right]$$

$$FV = 401.73 \cdot \left(1 + \frac{0.05}{4}\right) \left[\frac{\left(1 + \frac{0.05}{4}\right)^{4 \cdot 2} - 1}{\frac{0.05}{4}}\right] \approx 3400$$

$$3400 = PMT \cdot \left(1 + \frac{0.05}{4}\right) \left[\frac{\left(1 + \frac{0.05}{4}\right)^{4 \cdot 2} - 1}{\frac{0.05}{4}}\right] \quad PMT = 401.73$$

Explorations

Friday, April 26, 2019   2:15 AM

http://computerlearningservice.com/Academy/Finite-Math/Finance/Explorations/explorations.html

Compound Interest

Friday, April 26, 2019   2:17 AM

$$FV = PV \cdot \left(1 + \frac{r}{m}\right)^{m \cdot t} \qquad FV = 1000 \cdot \left(1 + \frac{0.04}{m}\right)^{m \cdot 10}$$

Rate	Period	Present Value	Future Value - 10 Years
4%	Annual - $m=1$	1000	1480.24
4%	Semiannual - $m=2$	1000	1485.95
4%	Quarterly - $m=4$	1000	1488.86
4%	Monthly - $m=12$	1000	1490.83
4%	Daily - $m=365$	1000	1491.79

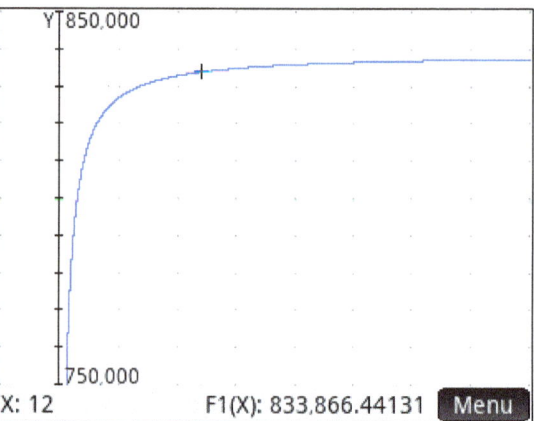

$$\lim_{x \to 0^+} 1000 \cdot \left(1 + \frac{0.04}{x}\right)^{x \cdot 10} = 1000$$

$$FV = PMT \left[\frac{\left(1 + \frac{r}{m}\right)^{m \cdot t} - 1}{\frac{r}{m}}\right] \qquad FV = 500 \left[\frac{\left(1 + \frac{0.0626}{12}\right)^{12 \cdot 10} - 1}{\frac{0.0625}{12}}\right] = 83060.94$$

$$FV_{\text{Annuity}} = PV_{\text{Loan}}$$

$$FV = F1(x) = 83060.94 \cdot \left(1 + \frac{0.0625}{x}\right)^{x \cdot 37}$$

Rate	Period	Present Value	Future Value - 37 Years
6.25%	Annual - $m=1$	83060.94	782649.97
6.25%	Semiannual - $m=2$	83060.94	809710.92
6.25%	Quarterly - $m=4$	83060.94	824019.47
6.25%	Monthly - $m=12$	83060.94	833866.44
6.25%	Daily - $m=365$	83060.94	838719.72

Continuous Compounding of Interest

Sunday, April 28, 2019   1:19 AM

n	$\left(1+\frac{1}{n}\right)^n$
1	2
10	2.5937424601
100	2.70481382942
1000	2.71692393223
10000	2.71814592672
∞	2.71828182846

Calculator display (Function, CAS):
- F1(1) = 2
- F1(10) = 2.5937424601
- F1(100) = 2.70481382942
- $\lim_{x \to 1000} (F1(x))$ = 2.71692393223
- $\lim_{x \to 10000} (F1(x))$ = 2.71814592672
- $\lim_{x \to \infty} (F1(x))$ = 2.71828182846

$$\lim_{n \to \infty}\left(1+\frac{1}{n}\right)^n = e \approx 2.71828$$

$$A = Pe^{rt} \qquad \text{contint(a,p,r,t):=ROUND(solve(a=p*(e\^{}(r*t)),x),2)}$$

Rate	Period	Present Value	Future Value - 10 Years
4%	Annual - m=1	1000	1480.24
4%	Semiannual - m=2	1000	1485.95
4%	Quarterly - m=4	1000	1488.86
4%	Monthly - m=12	1000	1490.83
4%	Daily - m=365	1000	1491.79
4%	Continuous	1000	1491.82

$$A = 1000 \cdot e^{0.04 \cdot 10}$$
$$= 1491.82$$

$$5000 = P \cdot e^{0.04 \cdot 10}$$
$$P = 3351.60$$

Subpage

Monday, April 29, 2019   4:18 AM

$$m = \left(1 + \frac{1}{1000}\right)^{1000}$$ Let *m* equal number

$$\ln m = \ln\left(1 + \frac{1}{1000}\right)^{1000}$$ Take ln of both sides

$$\ln m = 1000 \cdot \ln\left(1 + \frac{1}{1000}\right)$$ Power rule for ln

$$m = e^{1000 \cdot \ln\left(1 + \frac{1}{1000}\right)}$$ Definition of ln

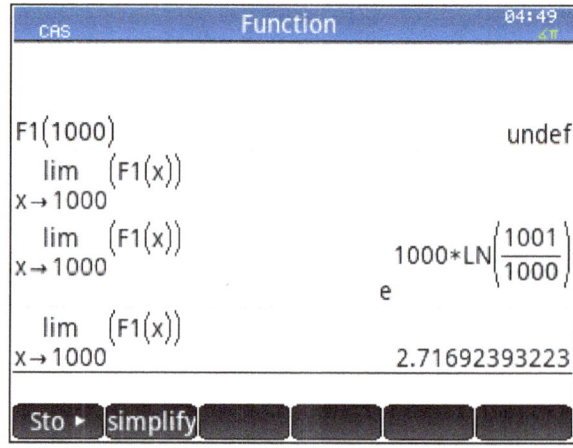

Effective Rate

Tuesday, April 30, 2019   3:46 AM

Rate	Period	Effective	Present Value	Future Value - 10 Years
4%	Annual - $m=1$	4%	1000	1480.24
4%	Semiannual - $m=2$	4.04%	1000	1485.95
4%	Quarterly - $m=4$	4.0604%	1000	1488.86
4%	Monthly - $m=12$	4.0742%	1000	1490.83
4%	Daily - $m=365$	4.0808%	1000	1491.7

$$FV = PV \cdot \left(1 + \frac{r}{m}\right)^{m \cdot t} \qquad r_{eff} = \left(1 + \frac{r}{m}\right)^m - 1$$

rateeff(f,r,m):=ROUND(solve(f=(1+r/m)^m-1,x),6)

$$r_{eff} = \left(1 + \frac{0.04}{365}\right)^{365} - 1$$

$$\approx 0.040808$$

$$A = 1000 \cdot (1 + 0{,}040808)^{10}$$

$$\approx 1491.78$$

$$0.0404 = \left(1 + \frac{r}{2}\right)^2 - 1$$

$$r = \frac{1}{25} \text{ or } r = -4.04$$

$$= \frac{1}{25} = 0.04$$

Sinking Fund

Monday, May 6, 2019   9:10 AM

A homeowner has decided to set up a sinking fund for the purpose of repairing and refinishing their kitchen cabinets in 2 years. The local shop has given then an estimate of $2200 for the doors and draws with an additional $1200 for a contractor to do the cabinet's frames. If the fund earns 5% per year compounded quarterly, determine the size of each quarterly installment.

$$FV = PMT \left[\frac{\left(1 + \frac{r}{m}\right)^{m \cdot t} - 1}{\frac{r}{m}} \right] \qquad A = P(1 + rt)$$

$$3400 = R \left[\frac{\left(1 + \frac{0.05}{4}\right)^{4 \cdot 2} - 1}{\frac{0.05}{4}} \right] \qquad R \approx 406.75$$

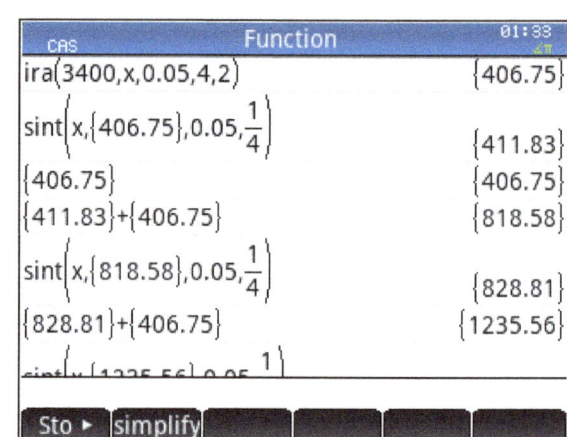

End of period	Installment	Interest Made	Addition	Amount in Fund
1	406.75	0	406.75	406.75
2	406.75	5.08	411.83	818.58
3	406.75	10.23	416.98	1235.56
4	406.75	15.44	422.19	1657.75
5	406.75	20.72	427.47	2085.22
6	406.75	26.07	432.82	2518.04
7	406.75	31.48	438.23	2956.27
8	406.75	36.95	443.70	3399.97

3

System of Equations and Matrices

> "The awkward moment is when you finish a math problem and your answer isn't even one of the choices."
> ~Ritu Ghatourey

Introductions – System of Linear Equations

- One Point   28
- Infinitely Many Solutions   29
- No Solution   30

Matrix Approach

- One Point   32
- First Element is a Zero   33
- Infinitely Many Solutions   34
- No Solution   35

Additional Investigations

- Tools 37
- Higher Equations Higher Unknowns 39
- Fewer Equations Than Unknowns 41
- More Equations Than Unknowns 43

Matrices

- Matrix Arithmetic 46
- Matrix Algebra 48
- Multiplication 50
- Inverse 51
    - ✓ Pivmat – Row Operations 52
    - ✓ Matrix Equation 53
    - ✓ Not Sufficient 54

Introductions - Systems of Linear Equations

Wednesday, May 29, 2019 1:11 AM

http://computerlearningservice.com/Academy/Finite-Math/SimEqs-Matrices/One-Point/one-point.html

One Point

Wednesday, May 29, 2019   1:16 AM

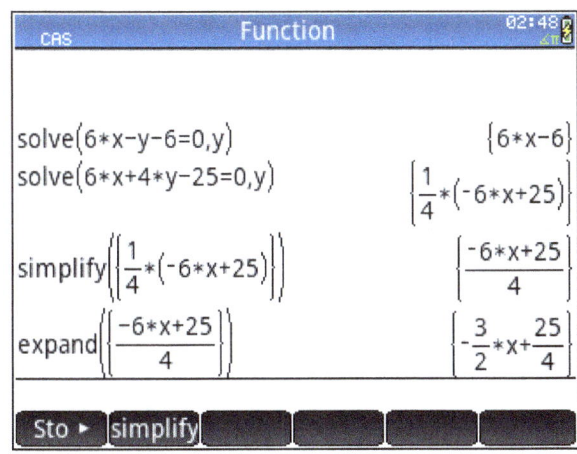

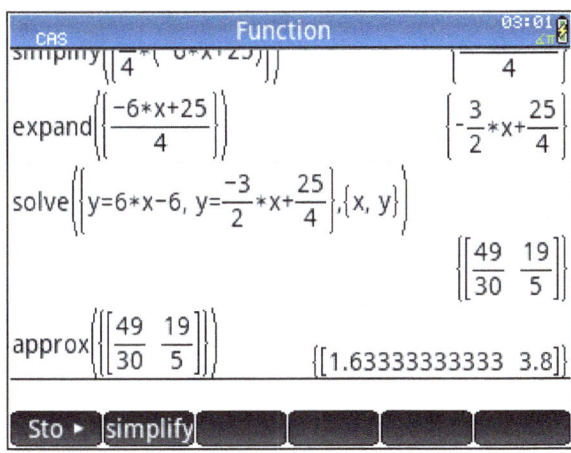

$$6x - 6 = -\frac{3}{2}x + \frac{25}{4}$$

Consistent System

$$x = \frac{49}{30} \text{ or } \left(\frac{49}{30}, \frac{19}{5}\right)$$

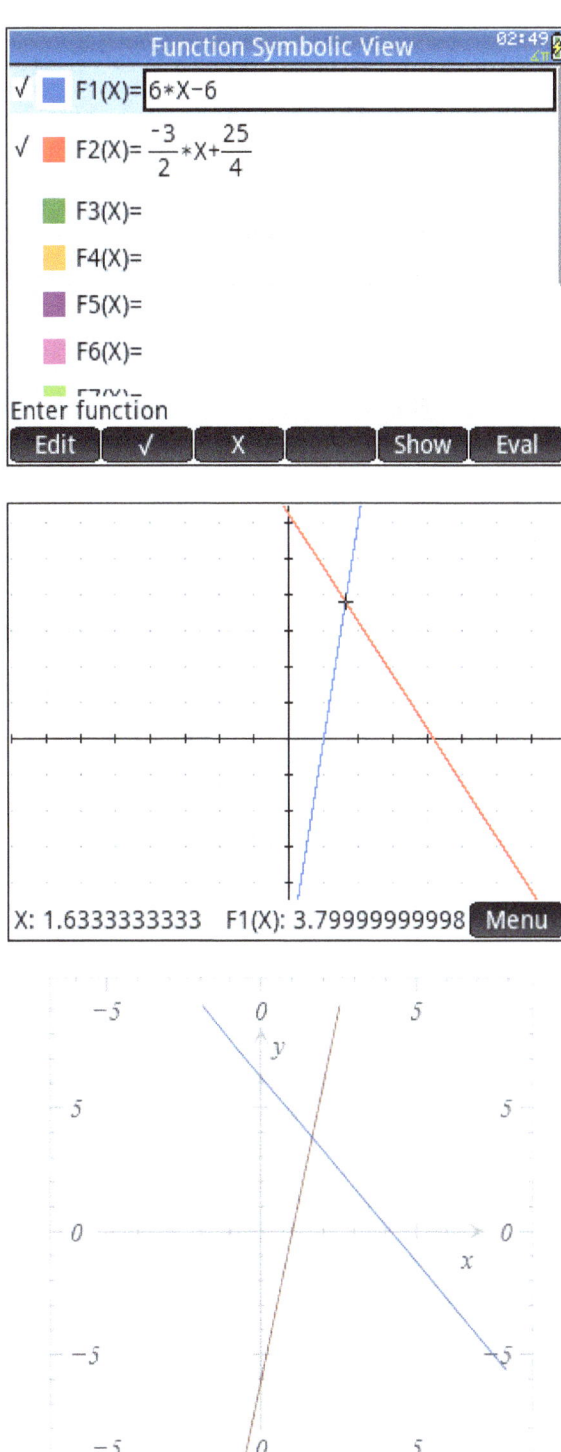

Infinitely Many Solutions

Wednesday, May 29, 2019   1:24 AM

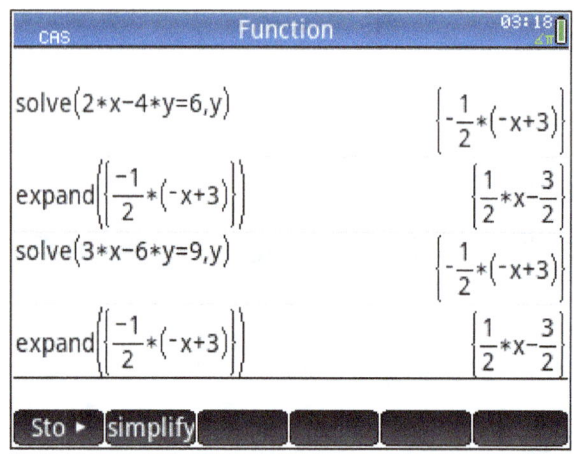

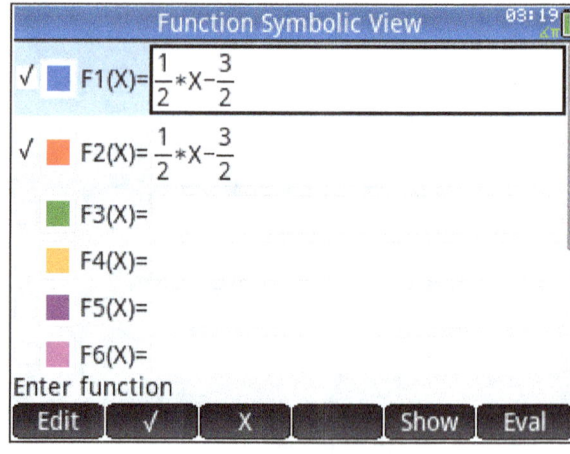

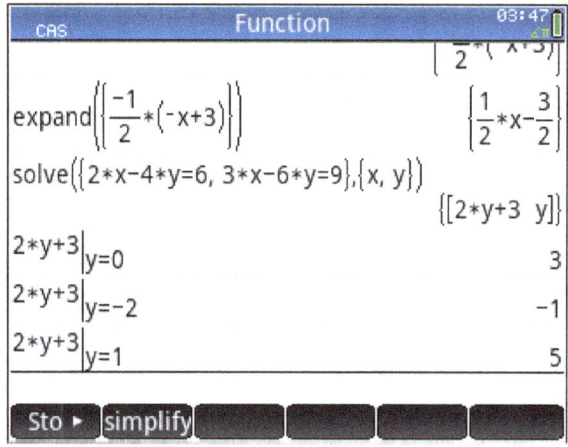

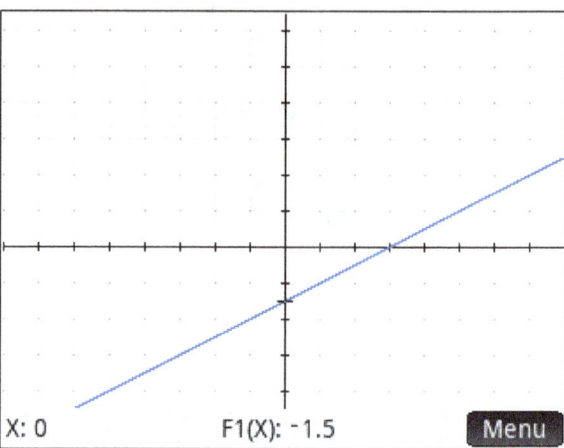

$$\frac{1}{2}x - \frac{3}{2} = \frac{1}{2}x - \frac{3}{2}$$

Dependent and Consistent System

$x \in \mathbb{R} \ or \ (2y + 3, y) \ or \ \left(x, \frac{1}{2}x - \frac{3}{2}\right)$

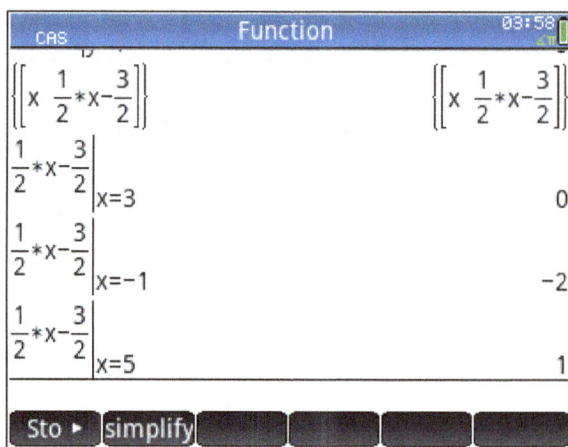

No Solution

Wednesday, May 29, 2019   1:25 AM

$3x + y = 1$
$3x + y = 4$

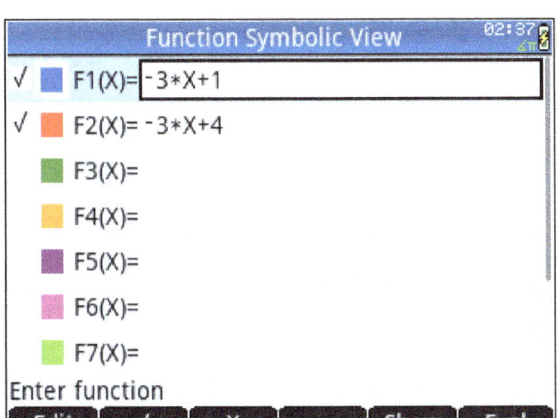

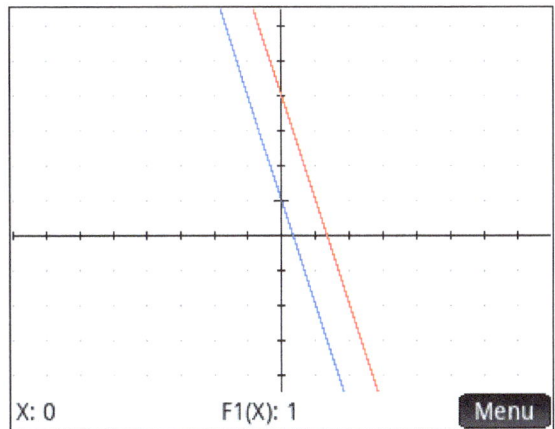

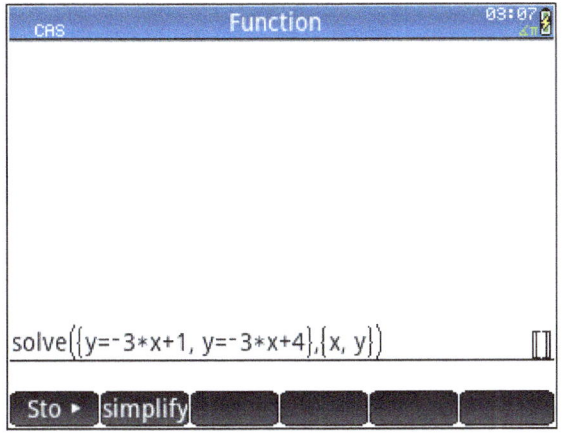

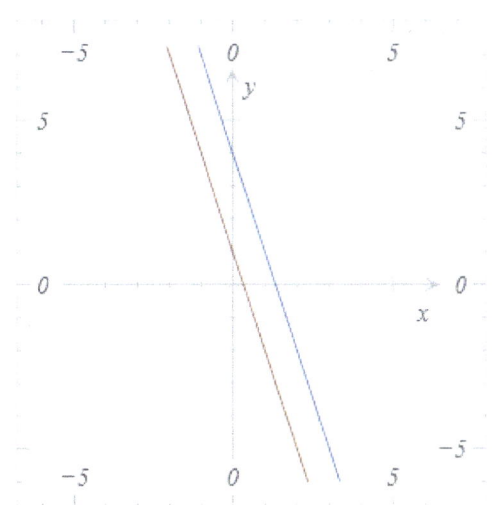

$$-3x + 1 = -3x + 4$$

Inconsistent System
$\emptyset$ or [ ]

Inconsistent System

Matrix Approach

Saturday, August 24, 2019   5:53 AM

http://computerlearningservice.com/Academy/Finite-Math/SimEqs-Matrices/Matrix-Approach/matrix-approach.html

One Point

Sunday, June 2, 2019   1:17 AM

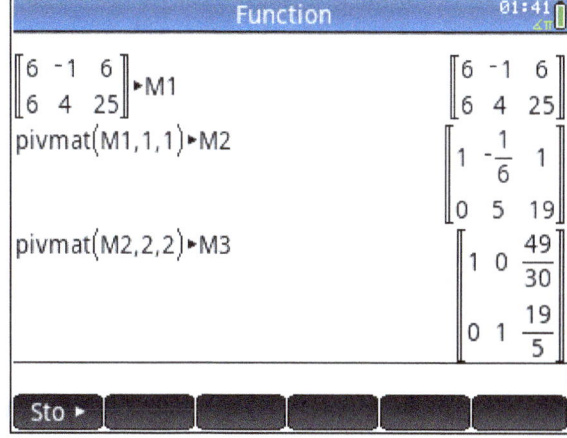

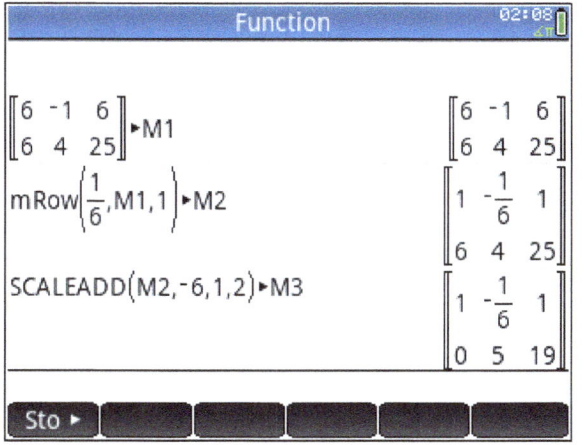

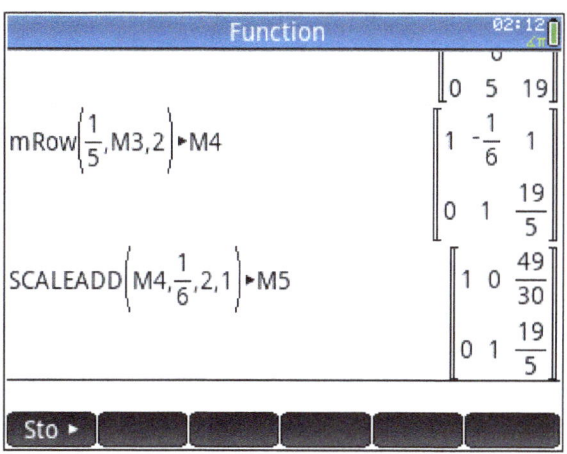

1. Any pivot zero replaced by non zero pivot [interchange rows - SWAPROWS]
2. Any row replaced by product of its reciprocal [pivot row - mRow]
3. Any non pivot row replaced by the sum(product(pivot row's times the opposite of the non pivot's column) and itself) [pivot column zero(s) - SCALEADD]

First Element is a Zero

Sunday, June 2, 2019   8:16 AM

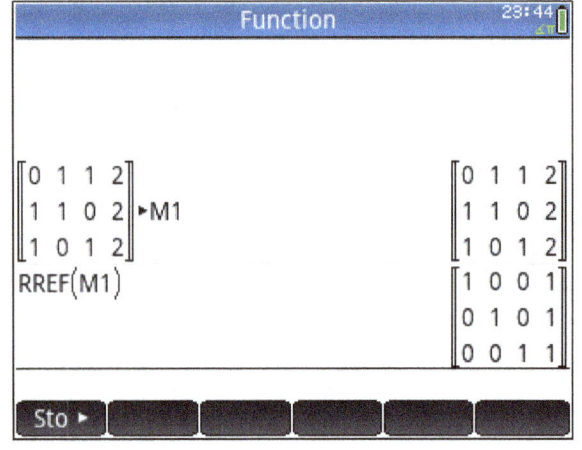
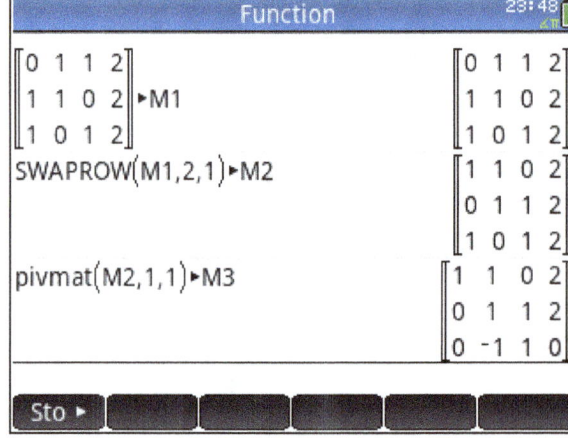
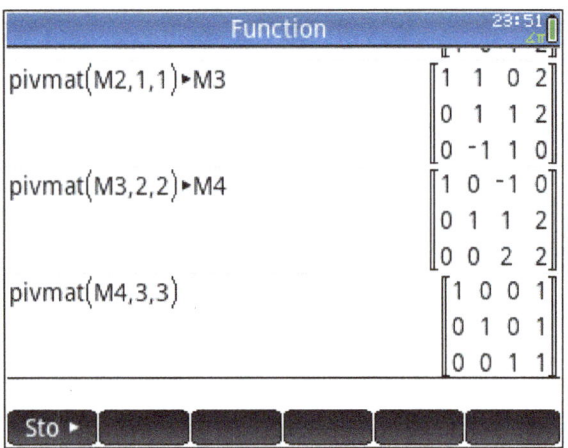
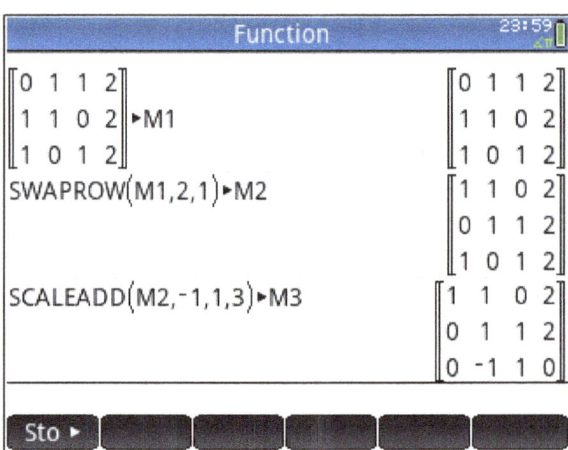
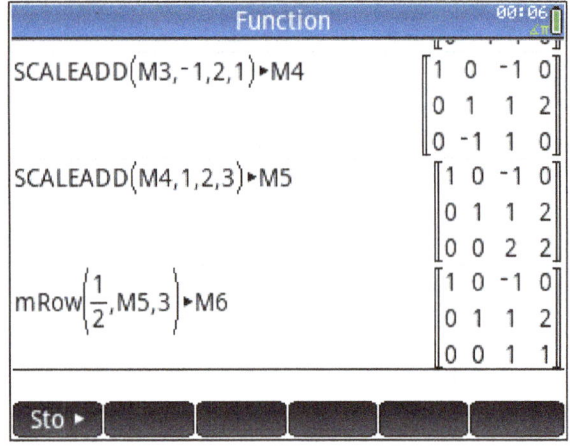
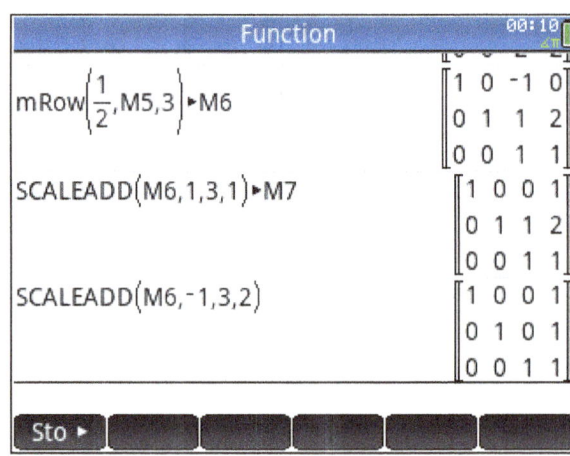

Infinitely Many Solutions

Monday, June 3, 2019   7:22 AM

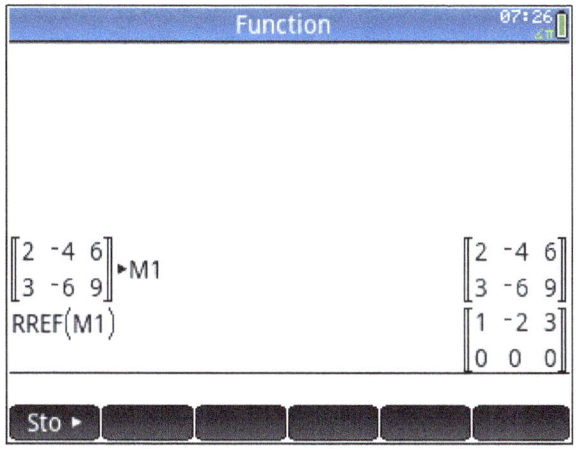

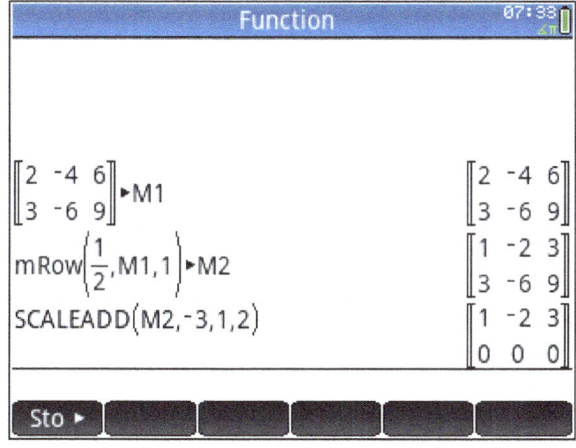

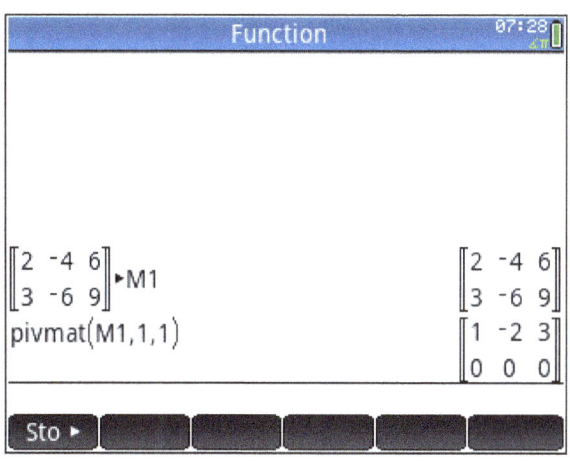

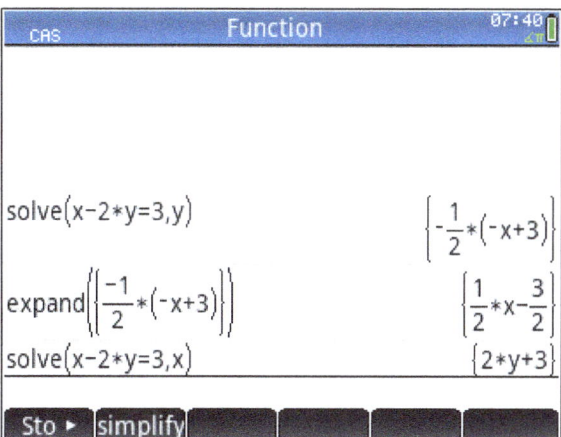

Dependent and Consistent System

$$x \in \mathbb{R} \;\; or \;\; (2y+3, y) \;\; or \;\; \left(x, \frac{1}{2}x - \frac{3}{2}\right)$$

1. Any pivot zero replaced by non zero pivot [interchange rows - SWAPROWS]
2. Any row replaced by product of its reciprocal [pivot row - mRow]
3. Any non pivot row replaced by the sum(product(pivot row's times the opposite of the non pivot's column) and itself) [pivot column zero(s) - SCALEADD]

No Solution

Monday, June 3, 2019   11:33 AM

$3x + y = 1$
$3x + y = 4$

Inconsistent System
$\emptyset$ or $[\,]$

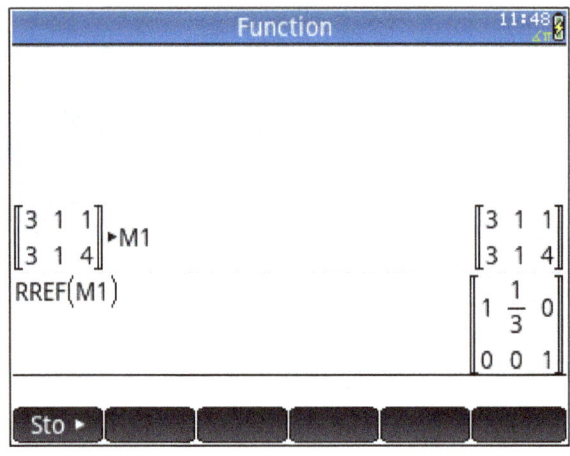

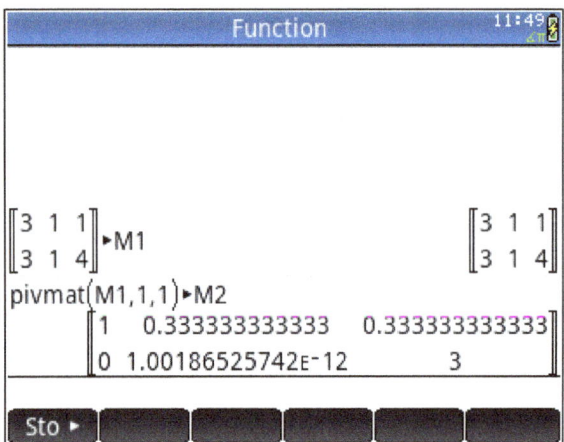

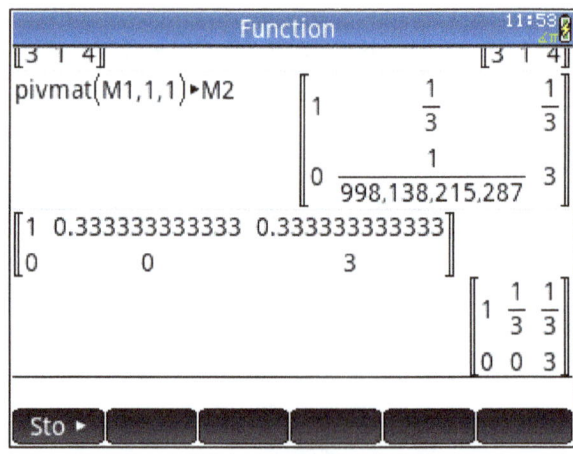

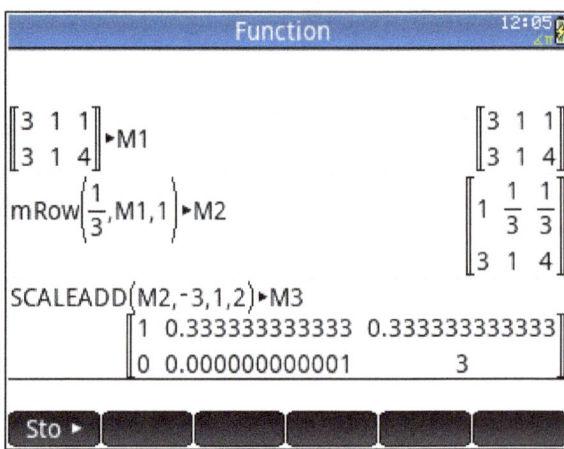

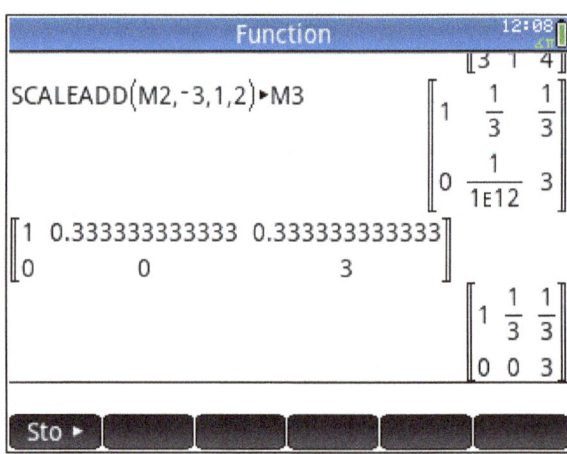

Additional Investigations

Tuesday, June 4, 2019   5:42 AM

http://computerlearningservice.com/Academy/Finite-Math/SimEqs-Matrices/Additional-Investigations/additional-investigations.html

Tools

Tuesday, June 4, 2019   5:43 AM

Systems	Row Operations	Pivmat	REF	RREF
2 equation 2 unknowns	x	x	x	x
3 equation 3 unknowns	x	x	x	x
Higher equations Higher unknowns	SWAPROW	x	x	x
Fewer equations than unknowns		x	x	x
More equations than unknowns		x	x	x

$3x + y = 1$

$3x + y = 4$

Inconsistent System

$\emptyset \; or \; [\;]$

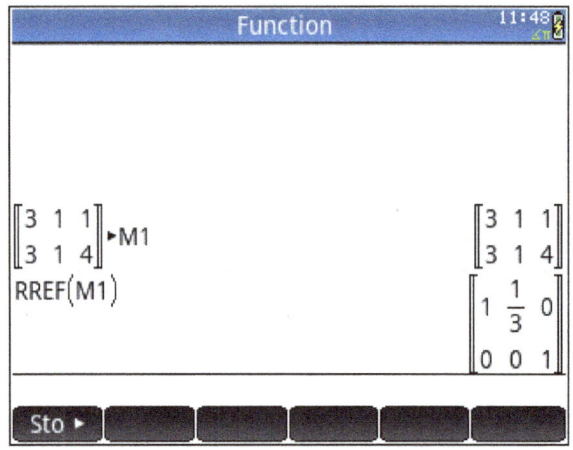

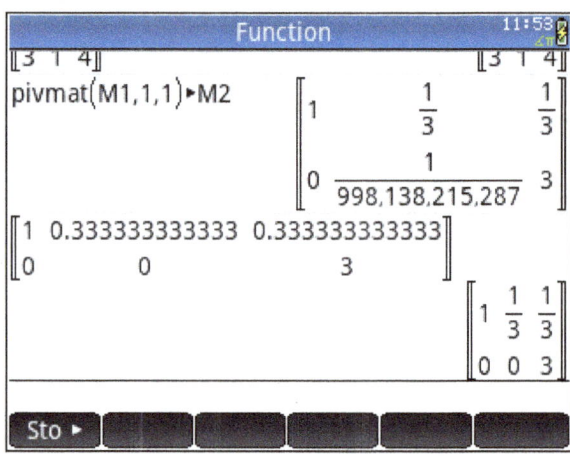

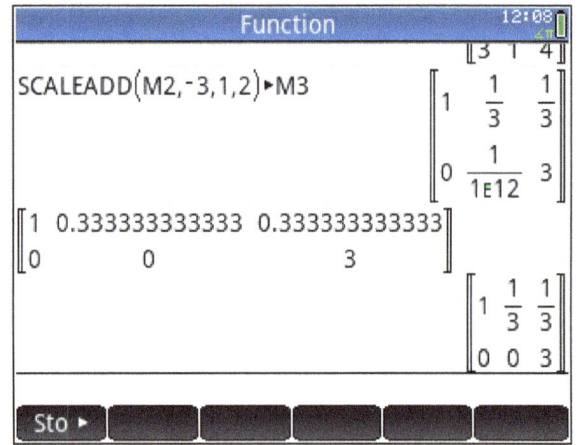

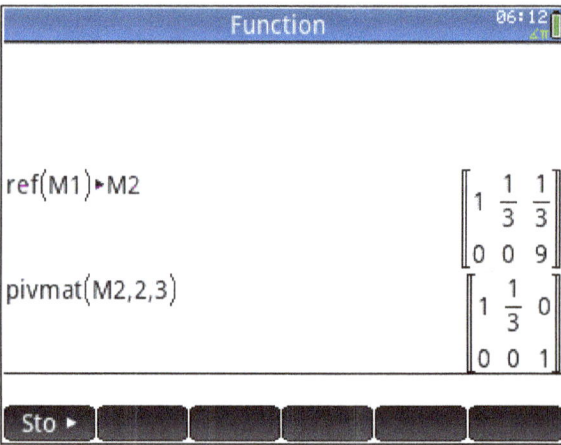

Subpage

Tuesday, June 4, 2019   6:14 AM

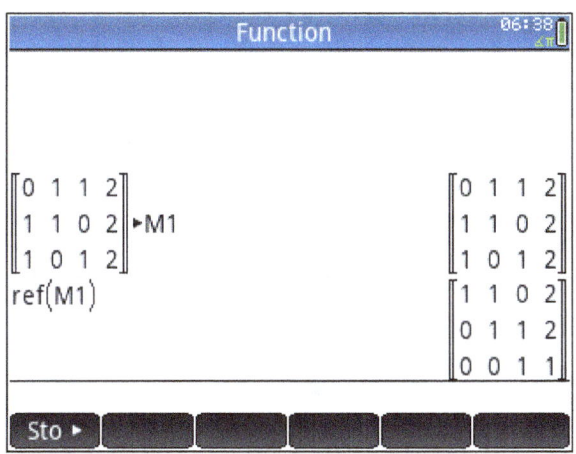
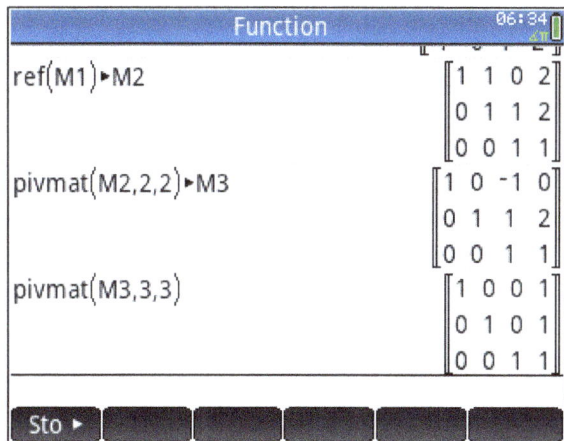

System of Equations and Matrices Page 38

Higher Equations Higher Unknowns

Tuesday, June 4, 2019   5:44 AM

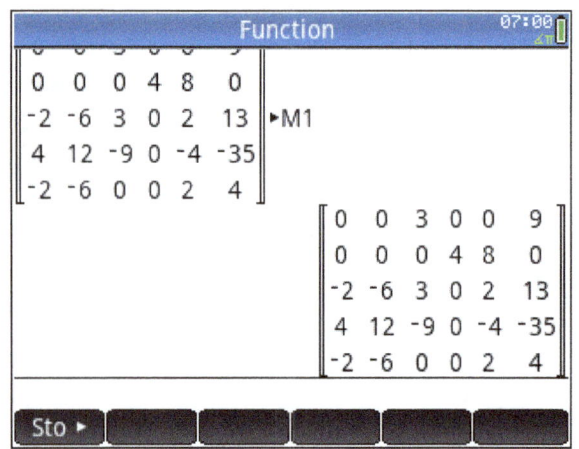
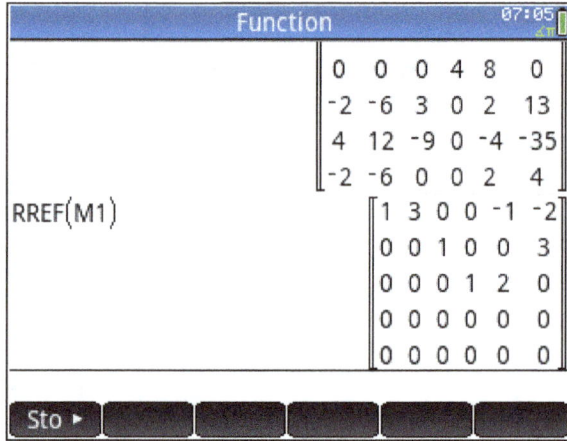
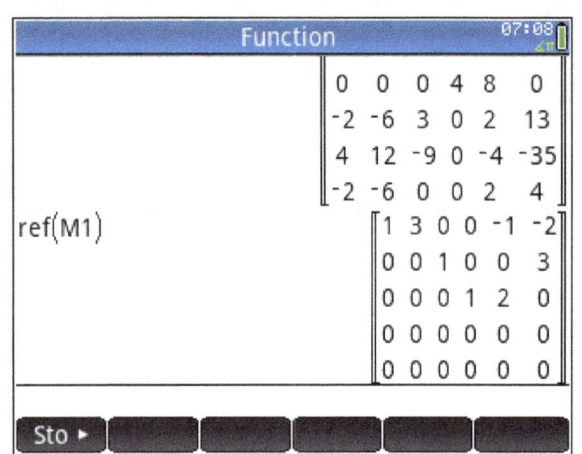
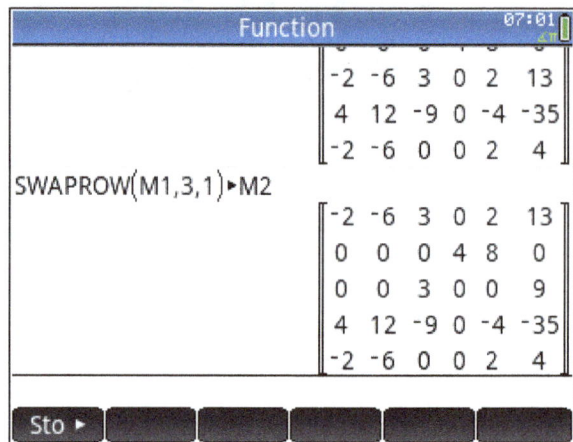
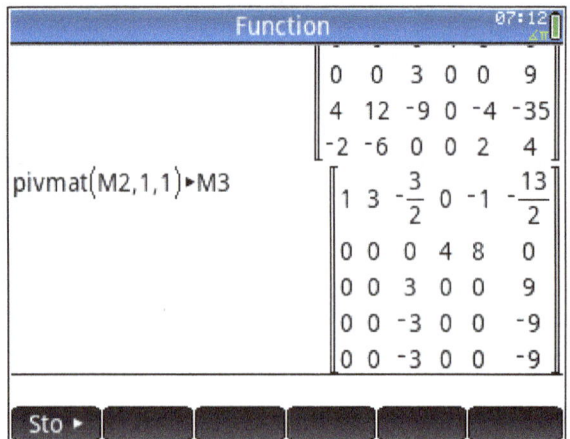
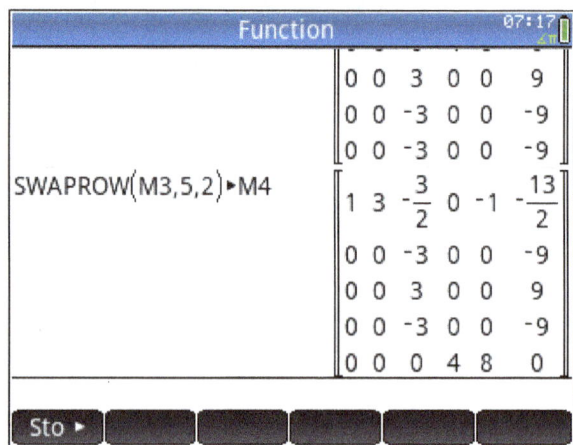

Subpage

Tuesday, June 4, 2019   7:19 AM

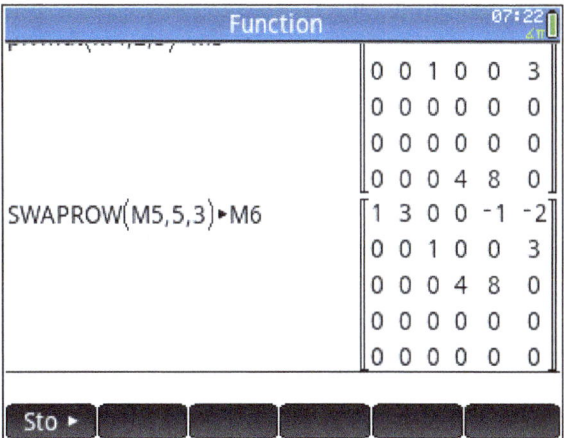
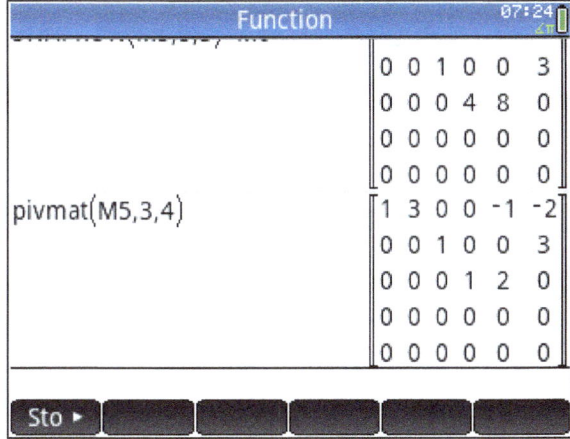

Dependent and Consistent System

$x_2 = s$

$x_5 = t$

$x_1 = -2 - 3s + t$

$x_3 = 3$

$x_4 = -2t$

Fewer Equations Than Unknowns

Tuesday, June 4, 2019   5:44 AM

$9x_1 + 12x_2 + 5x_3 = 100$

$3x_1 + 4x_2 + 4x_3 = 40$

Row Reduced Echelon Form
1. As we move from left to right, first nonzero of every row is 1.
2. Column with left most 1 of a row, all remaining entries are zeros.
3. Left 1 of any row is to right of left most 1 of preceding row.
4. Rows with only zeros are below the other rows.

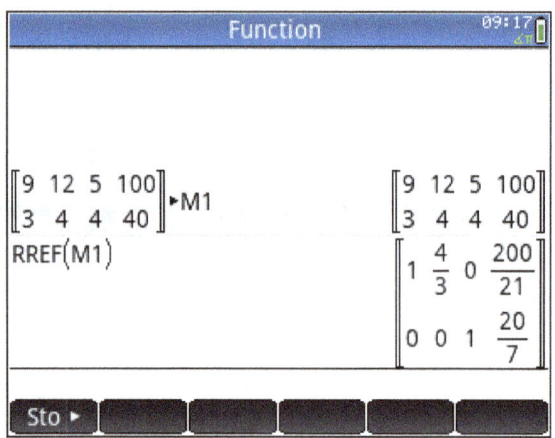
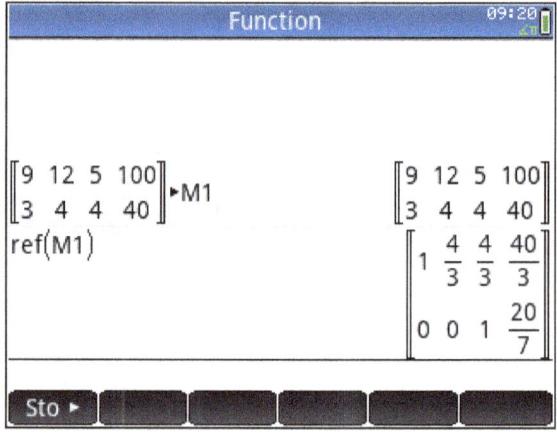
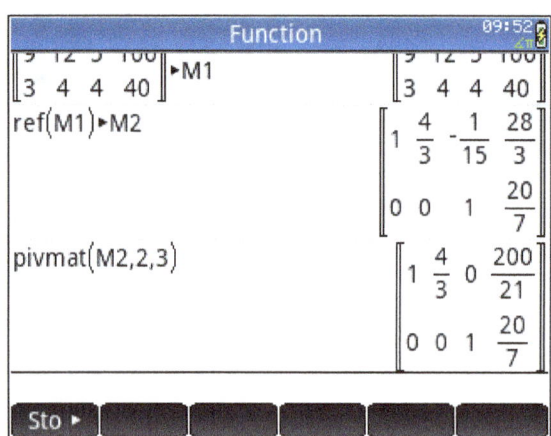
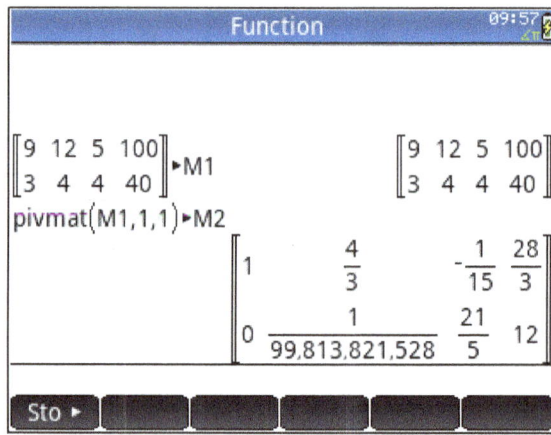

System of Equations and Matrices Page 41

Subpage

Tuesday, June 4, 2019   10:04 AM

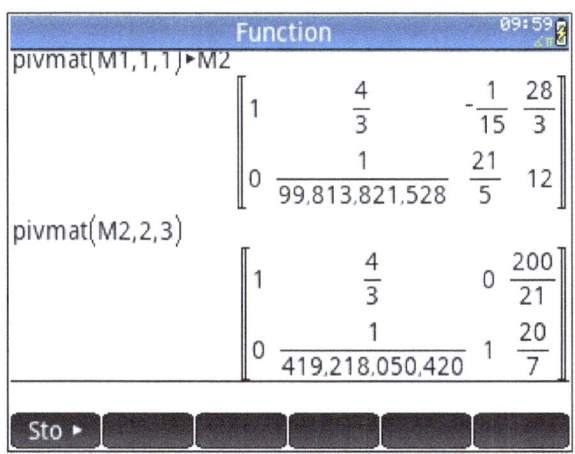

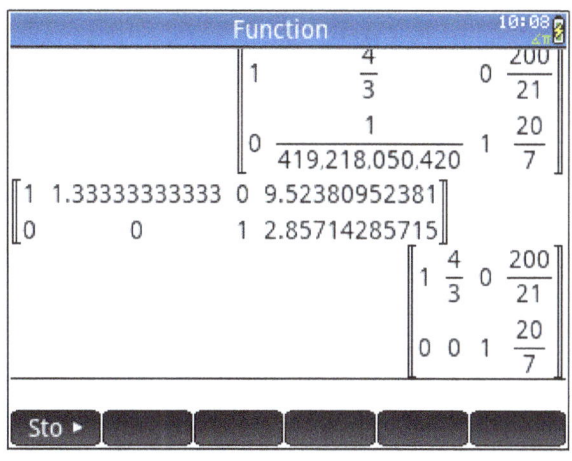

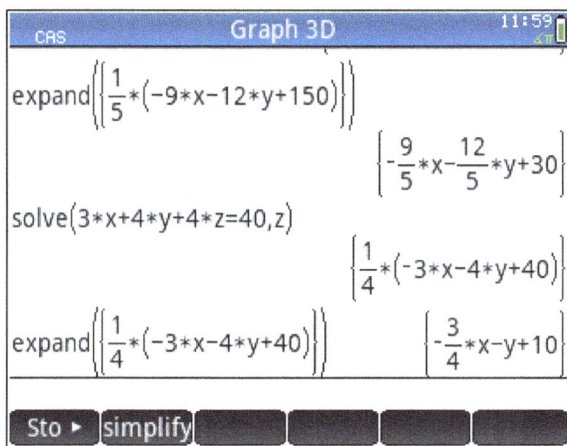

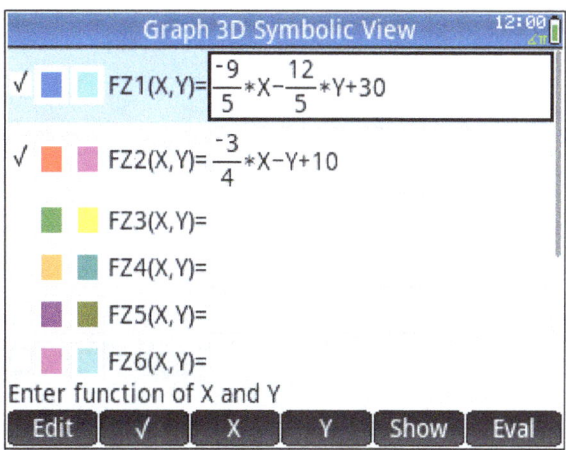

Dependent and Consistent System

$x_2 = s$

$x_1 = \dfrac{200}{21} - \dfrac{4}{3}s$

$x_3 = \dfrac{20}{7}$

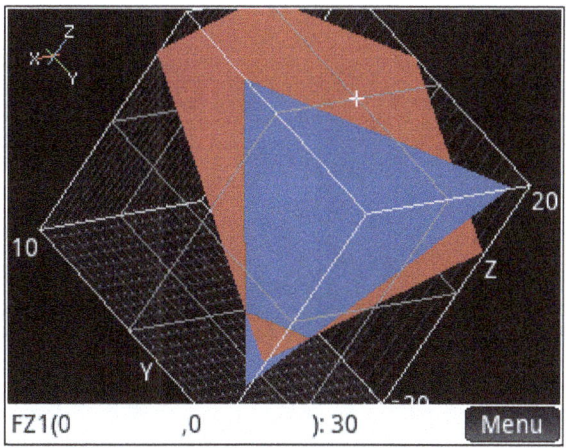

More Equations Than Unknowns

Tuesday, June 4, 2019   5:45 AM

Inconsistent System

$13x + 7y = 90$

$5x = 20$

$7x + 8y = 53$

$11x + 4y = 62$

$\emptyset \text{ or } [\,]$

Row Reduced Echelon Form

1. As we move from left to right, first nonzero of every row is 1.
2. Column with left most 1 of a row, all remaining entries are zeros.
3. Left 1 of any row is to right of left most 1 of preceding row.
4. Rows with only zeros are below the other rows.

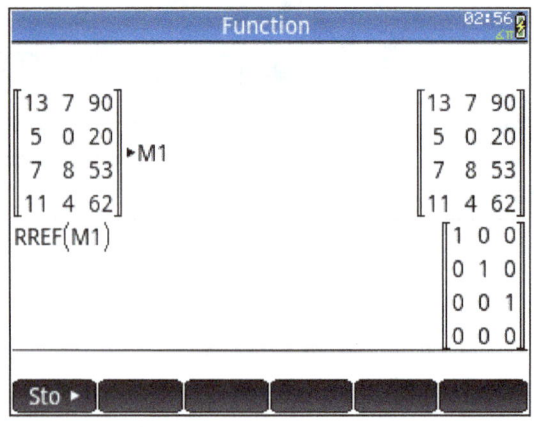

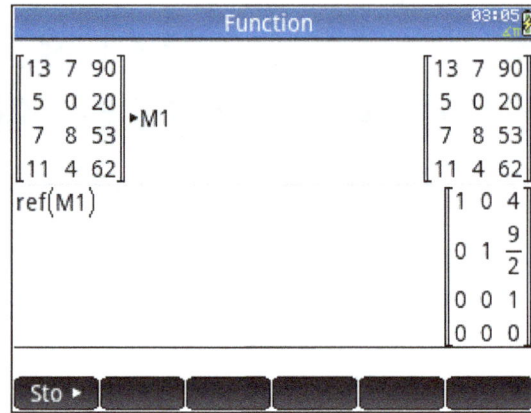

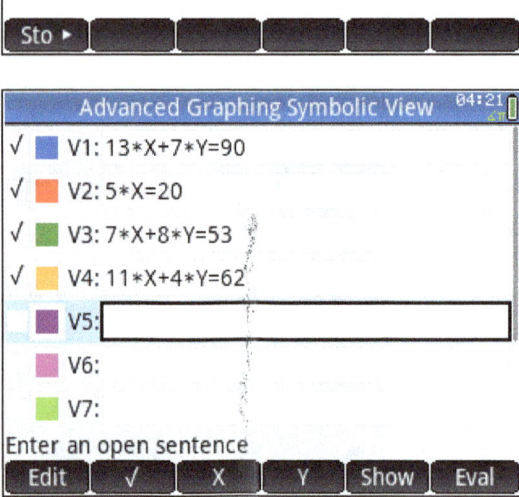

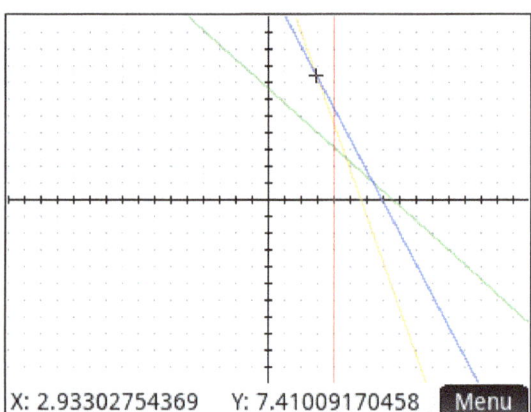

Subpage

Wednesday, June 5, 2019   3:57 AM

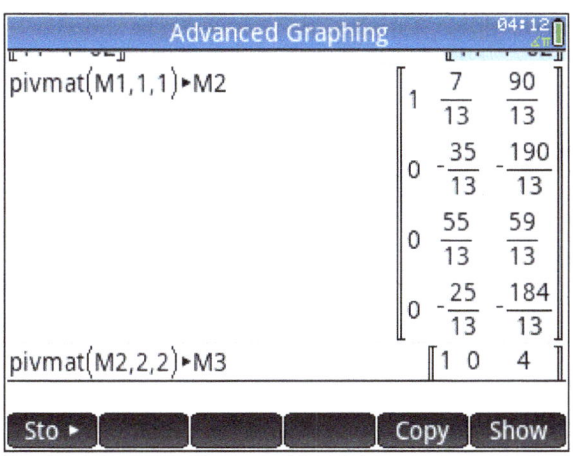

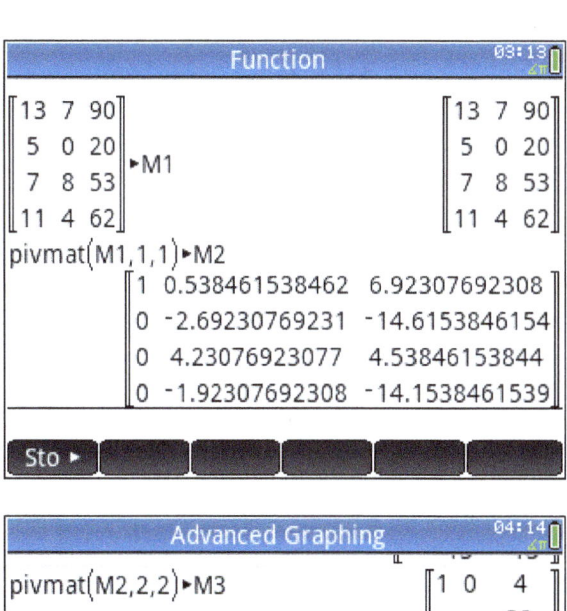

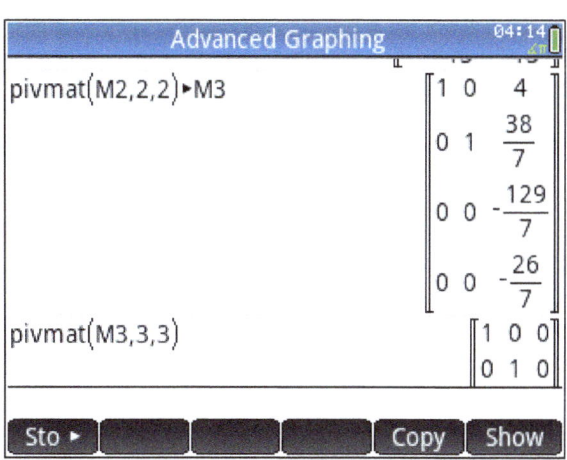

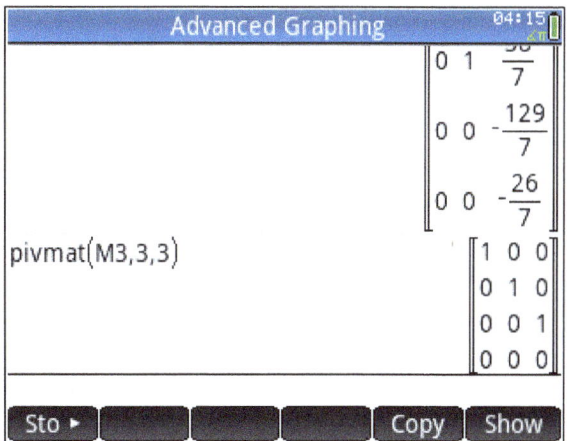

System of Equations and Matrices Page 44

Matrices

Saturday, June 15, 2019   8:24 AM

http://computerlearningservice.com/Academy/Finite-Math/SimEqs-Matrices/Matrices/matrices.html

Matrix Arithmetic

Saturday, June 15, 2019   8:28 AM

Daily Production

Refrigerator	City A	City B
Basic	10	8
Standard	20	19
Deluxe	5	4

How many refrigerators produced in a day at City B?
How many standard models are produced in a day?

$$sumcol(m, c) := \Sigma\text{LIST}(col(m, c))$$

$$sumrow(m, r) := \Sigma\text{LIST}(row(m, r))$$

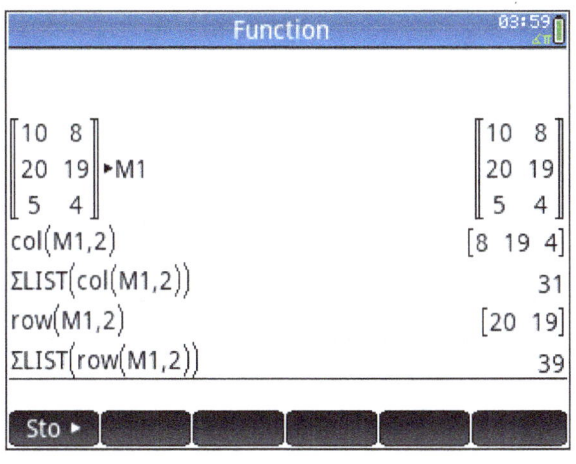

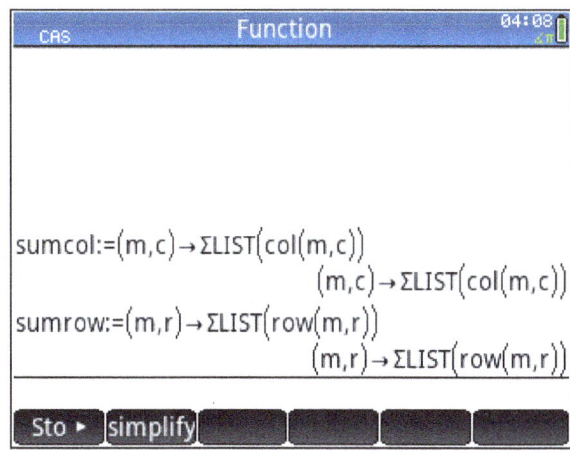

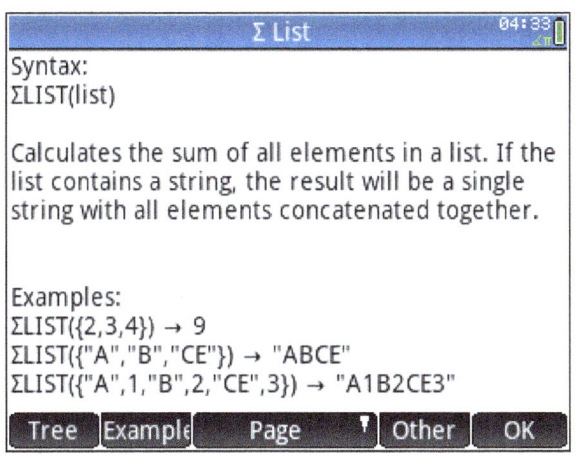

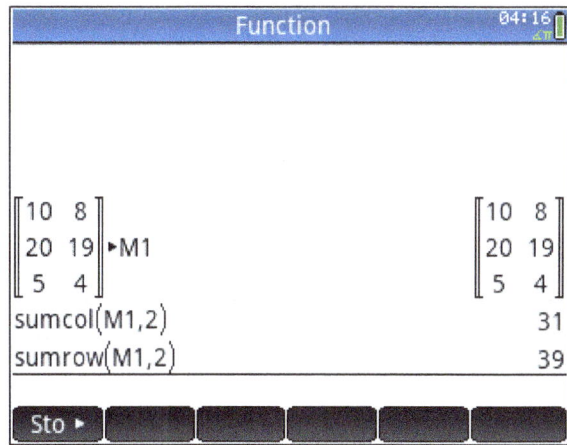

Subpage

Tuesday, June 18, 2019   4:17 AM

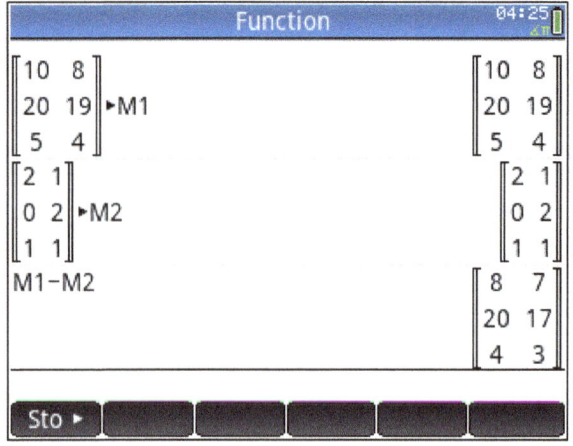

How many produced in June (assume 22 work days)?
Assume 8 hour day, with 2 hour overtime, daily production is?

A factory outlet is in City C. If they get the daily production refrigerators for June 15, shown by matrix M2, what amount is left for distribution channels?

Matrix Algebra

Saturday, June 15, 2019   4:10 PM

Properties of Matrices

$A + B = B + A$	Commutative addition
$(A + B) + C = A + (B + C)$	Associative addition
$A + 0 = 0 + A = A$	Additive identity
$A + (-A) = (-A) + A = 0$	Additive inverse
$A - B = A + (-B)$	Subtraction
kA	Scalar multiply (every element multiplied by k)
$k(A + B) = kA + kB$	Scalar distributive with matrices
$(k + l)A = kA + lA$	Scalar distributive with a matrix
$(kl)A = k(lA)$	Scalar associative multiplication

Refer to the following matrices:

$$A = \begin{bmatrix} 3 & -1 \\ -3 & 4 \\ 0 & 2 \end{bmatrix} \quad B = \begin{bmatrix} -4 & 2 \\ 1 & 0 \\ 2 & -3 \end{bmatrix}$$

$$C = \begin{bmatrix} -4 & 3 \\ 0 & -1 \\ 5 & 2 \end{bmatrix} \quad D = \begin{bmatrix} -2 & 1 \\ 0 & -3 \end{bmatrix}$$

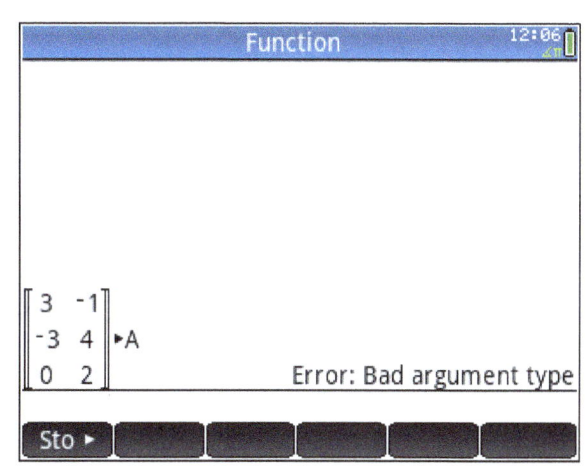

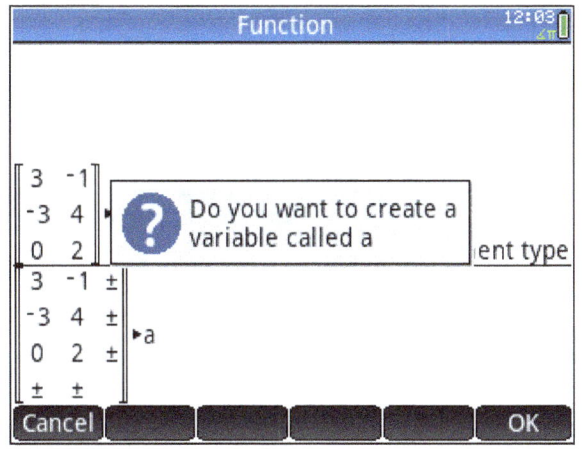

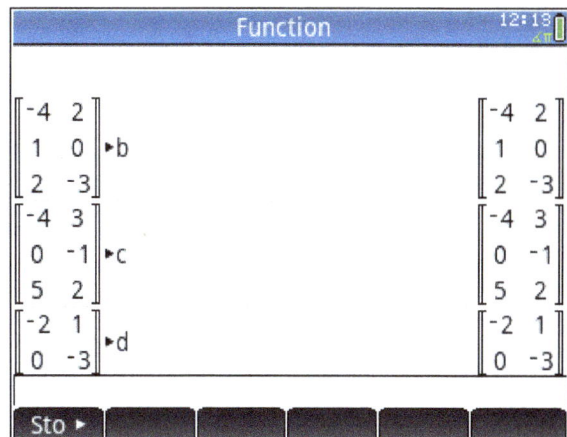

System of Equations and Matrices Page 48

Subpage

Tuesday, June 18, 2019   12:15 PM

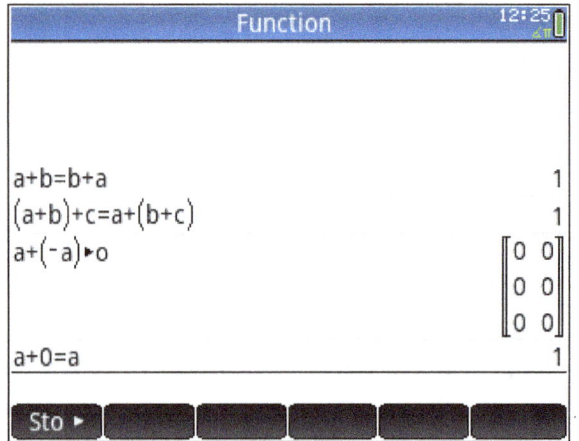
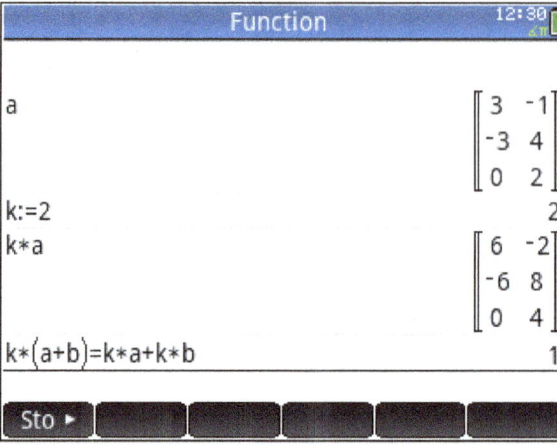

$A + B = B + A$	Commutative addition
$(A + B) + C = A + (B + C)$	Associative addition
$A + 0 = 0 + A = A$	Additive identity
$A + (-A) = (-A) + A = 0$	Additive inverse
kA	Scalar distributive with matrices

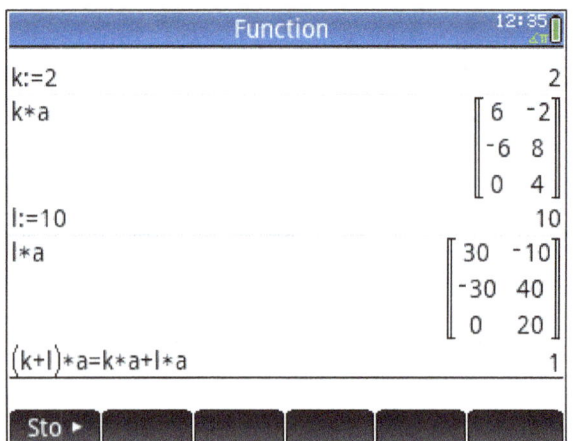
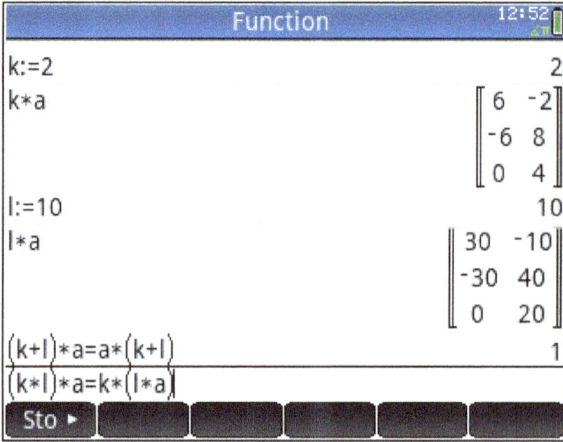

$k(A + B) = kA + kB$	Scalar distributive with matrices
$(k + l)A = kA + lA$	Scalar distributive with a matrix
$(kl)A = k(lA)$	Scalar associative multiplication

Multiplication

Sunday, June 16, 2019   3:13 AM

$$\underset{(2\times 3)}{\begin{bmatrix} a_{11} & a_{12} & a_{13} \\ a_{21} & a_{22} & a_{23} \end{bmatrix}} \underset{(3\times 4)}{\begin{bmatrix} b_{11} & b_{12} & b_{13} & b_{14} \\ b_{21} & b_{22} & b_{23} & b_{23} \\ b_{31} & b_{32} & b_{33} & b_{34} \end{bmatrix}} = \underset{(2\times 4)}{\begin{bmatrix} r_1c_1 & r_1c_2 & r_1c_3 & r_1c_4 \\ r_2c_1 & r_2c_2 & r_2c_3 & r_2c_4 \end{bmatrix}}$$

$$\begin{bmatrix} (a_{11}\cdot b_{11} + a_{12}\cdot b_{21} + a_{13}\cdot b_{31}) & (a_{11}\cdot b_{12} + a_{12}\cdot b_{22} + a_{13}\cdot b_{32}) & \ldots & \ldots \\ (a_{21}\cdot b_{11} + a_{22}\cdot b_{21} + a_{23}\cdot b_{31}) & (a_{21}\cdot b_{12} + a_{22}\cdot b_{22} + a_{23}\cdot b_{32}) & \ldots & \ldots \end{bmatrix}$$

$$\begin{bmatrix} (\text{row 1}\cdot\text{column 1}) & (\text{row 1}\cdot\text{column 2}) & (\text{row 1}\cdot\text{column 3}) & (\text{row 1}\cdot\text{column 4}) \\ (\text{row 2}\cdot\text{column 1}) & (\text{row 2}\cdot\text{column 2}) & (\text{row 2}\cdot\text{column 3}) & (\text{row 2}\cdot\text{column 4}) \end{bmatrix}$$

This means the sum of all the [(respective elements in a row)·(all the respective elements in the column)] is a single element in the result matrix.

Contract for Refrigerators

	Basic	Standard	Deluxe
Retailer 1	800	1000	1100
Retailer 2	790	1050	1075

Orders for Factory in City

	City A	City B	City C	City D
Basic	70	70	75	80
Standard	80	90	100	110
Deluxe	30	35	40	45

Retailer Paid Factory in City

	City A	City B	City C	City D
Retailer 1	169,000	184,500	204,000	223,500
Retailer 2	171,550	187,425	207,250	227,075

Multiplicative Properties of Matrices

$(AB)C = A(BC)$	Associative multiplication
$A(B + C) = AB + AC$	Left distributive with matrices
$(B + C)A = BA + CA$	Right distributive with matrices

$$indelem(a, b, n, m) := \Sigma\text{LIST}(row(a,n) .* col(b,m))$$

Inverse

Sunday, June 16, 2019   7:50 AM

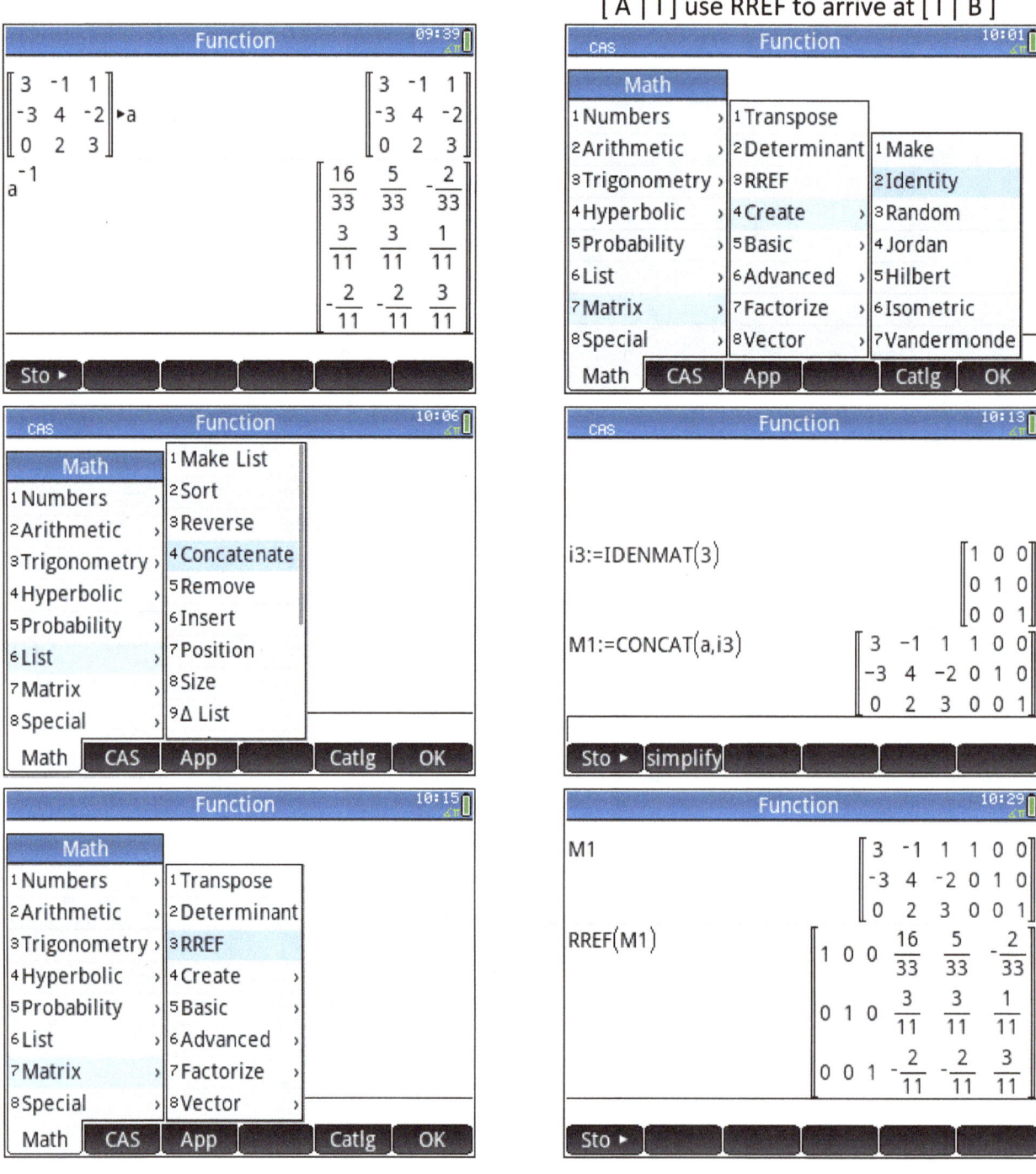

Pivmat - Row Operations

Tuesday, June 18, 2019   10:20 AM

[A | I] use pivmat or row operations to arrive at [I | B]

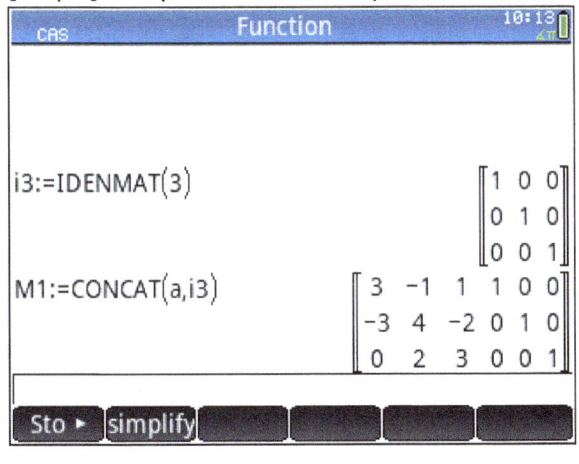

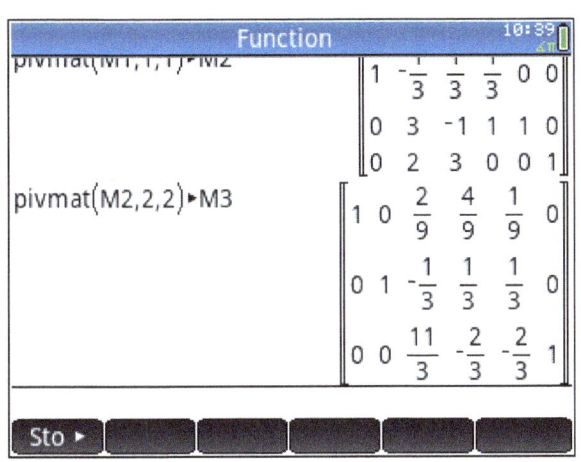

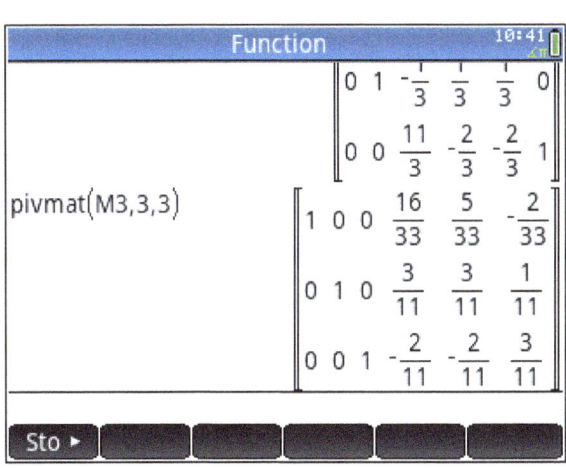

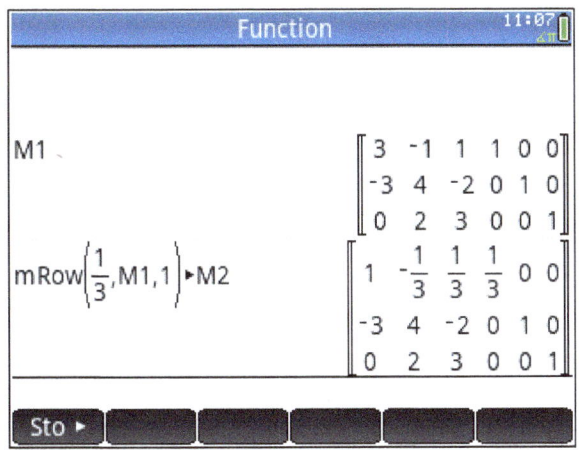

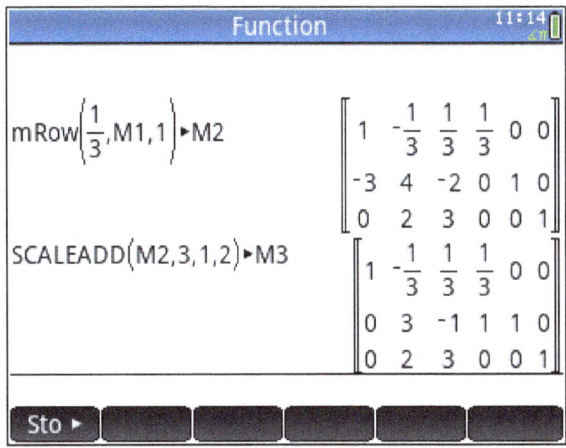

Matrix Equation

Sunday, June 16, 2019   9:47 AM

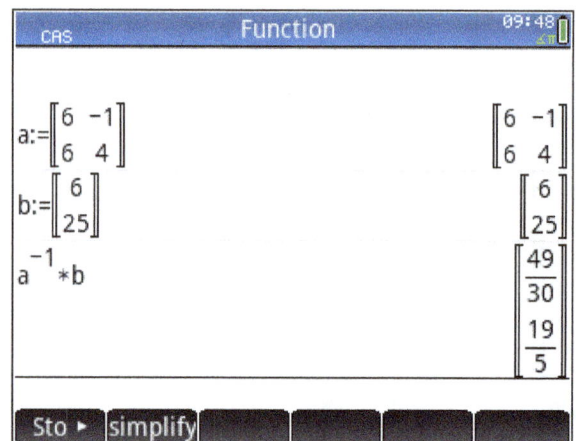

$AX = B$ Given
$A^{-1}(AX) = A^{-1}B$ Multiplication both sides
$(A^{-1}A)X = A^{-1}B$ Associative multiplication
$IX = A^{-1}B$ Inverse
$X = A^{-1}B$ Identity

Steps for Solving Linear Equation

- The equation is in standard form.
$AX = B$

- Divide both sides by A.
$$\frac{AX}{A} = \frac{B}{A}$$

- Dividing by A undoes the multiplication by A.
$$X = \frac{B}{A}$$

Not Sufficient

Tuesday, June 18, 2019   10:19 AM

Using the Matrix Editor

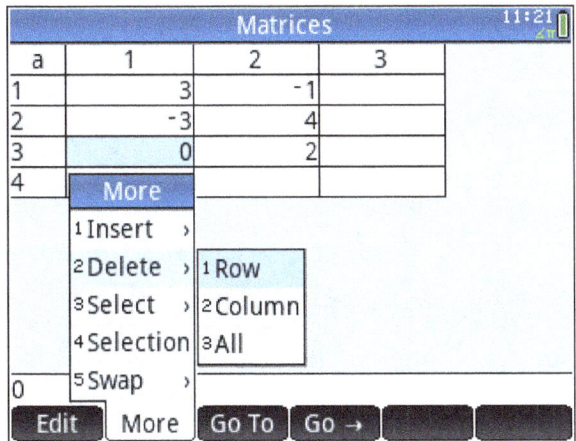
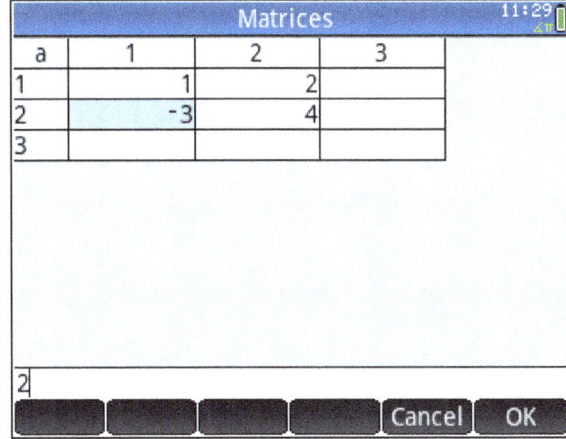

Necessary to be square but not sufficient

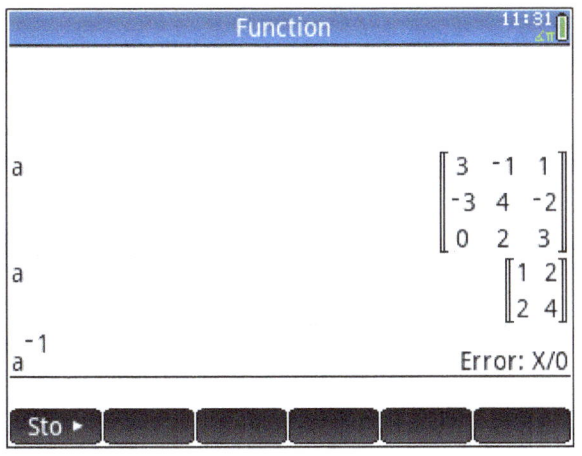
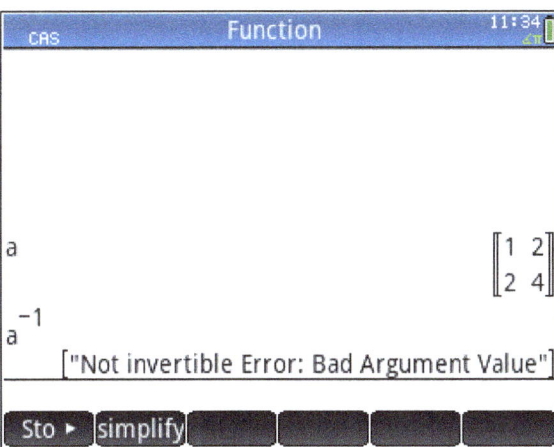

System of Equations and Matrices Page 54

4

Linear Programing

> "Just because we can't find a solution it
> doesn't mean that there isn't one."
> ~ Andrew Wiles

Graphical Maximums

- Bounded – Standard   56
- Bounded – Nonstandard   58

Graphical Minimums

- Unbounded – Standard   61
    - ✓ Intercepts   62
    - ✓ Simultaneous Equations   63
- Line – Nonstandard   64

Simplex Maximums

- Standard   67
- Nonstandard   69

Simplex Minimums

- The Dual   72
- Line – Nonstandard   74

Graphical Maximums

Sunday, April 21, 2019　3:44 PM

http://computerlearningservice.com/Academy/Finite-Math/Dual-Problem/Graphical-Maximums/graphical-maximums.html

Bounded - Standard

Monday, May 13, 2019   12:30 PM

To satisfy stockholders a company needs to make a profit of $5 on each goose down pillow and $9 on each goose down comforter it manufactures. The company buys high-quality goose down from a distributor who guarantee a supply of 2,500 pounds a week. The factory is limited to 45 workers at 40 hours per week. The pillow requires 2 pounds of down, 1 hour of work and comforter 4 pounds of down, 3 hours of work. How many of each per week should the company manufacture for maximum profit?

	Per pillow	Per comforter	Total
Down	2 pounds	4 pounds	2500 pounds
Time	1 hours	3 hours	1800 hours
Profit	$5	$9	

Maximize $P = 5x + 9y$

Subject to $x \geq 0, y \geq 0$

$$2x + 4y \leq 2500$$
$$x + 3y \leq 1800$$

Corners	(0,0)	(1250,0)	(150,550)	(0,600)
	0	6250	5700	5400

The company should make 1,250 goose down pillows per week for a profit of $6,250.

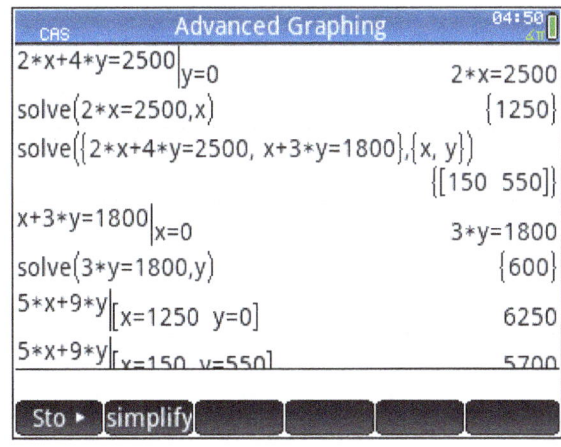

Subpage

Tuesday, May 14, 2019   3:55 AM

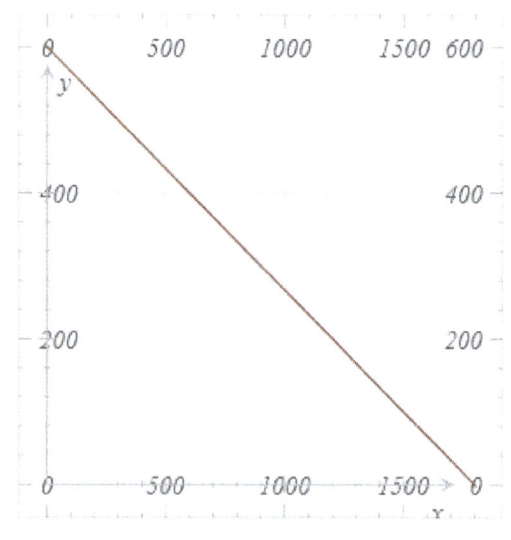

$2x + 4y \leq 2500$

$x + 3y \leq 1800$

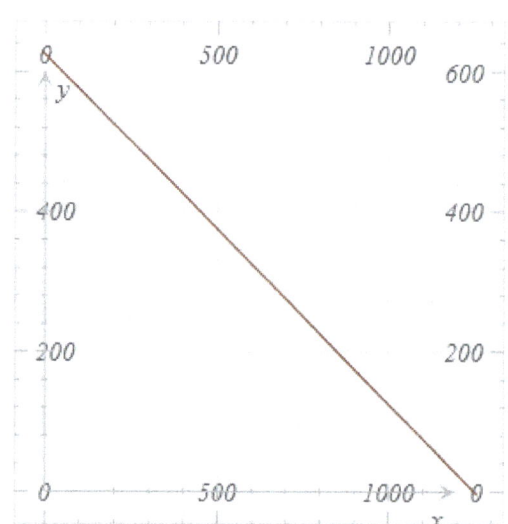

(0,600), (1800,0); (0,600), (1300,0)
Range [2000 by 600] Tick [500 by 200]

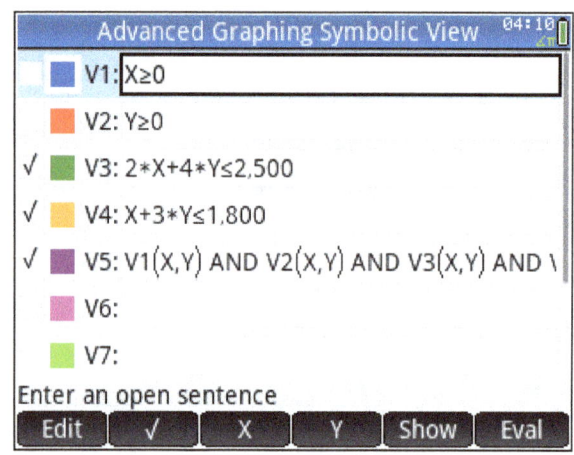

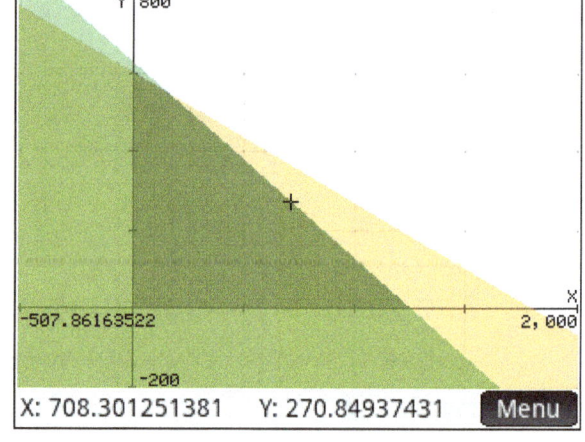

| Corners | (0,0) | (1250,0) | (150,550) | (0,600) |

Bounded - Nonstandard

Monday, May 20, 2019   7:06 AM

Maximize $P = 3x - y$
Subject to $x \geq 0, y \geq 0$
            $5x - y \geq 2$
            $x - 5y \leq 18$
            $x + y \leq 12$

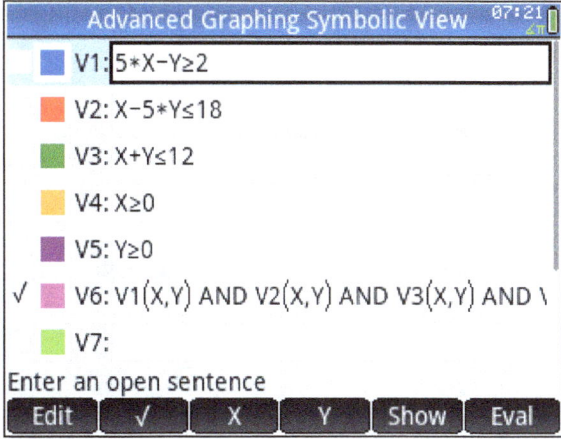

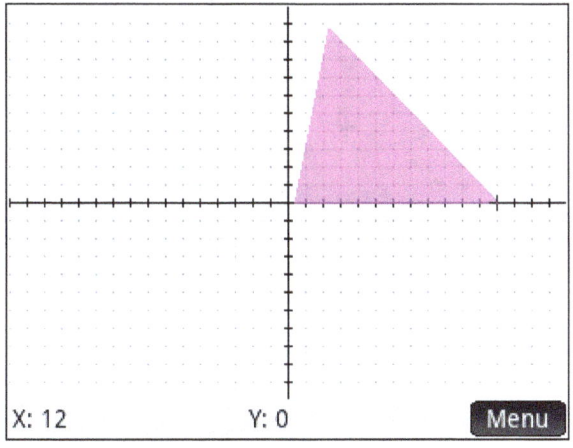

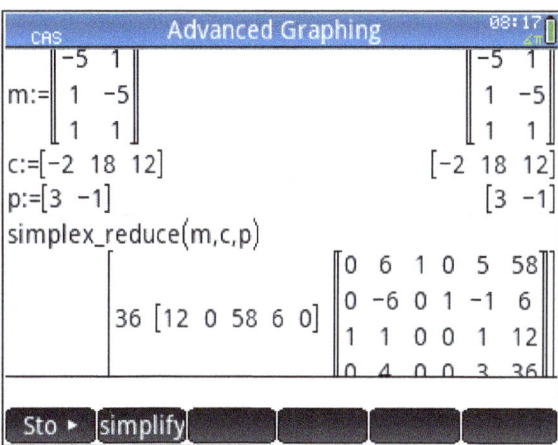

Corners	(2/5, 0)	(12, 0)	(7/3, 29/3)
p(x,y)	6/5	36	-8/3

Maximum of 36 at x=12 and y=0. Note simplex_reduce correct answer though it is not applicable.

simplex_reduce(Matrix_A, Vector_B, Vector_C) is not applicable. $5x - y \geq 2 \rightarrow -5x + y \leq -2$

Simplex Reduction help specifies:
 A_x <= 0 and x>=0, b>= 0. c(1,1)=-2 violates b>=0.

Subpage

Monday, May 20, 2019   7:25 AM

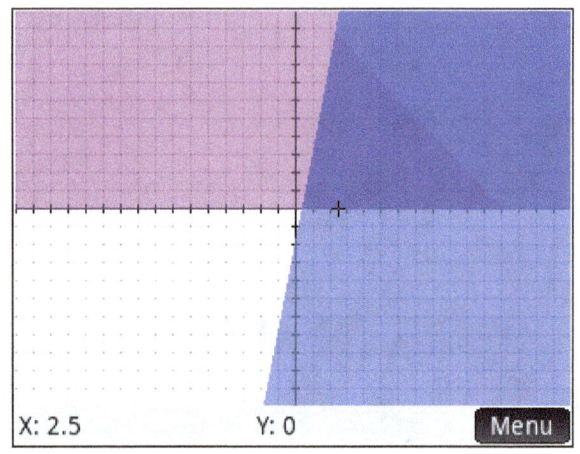

| $5x - y \geq 2$ | $y \geq 0$ | (5/2, 0) |

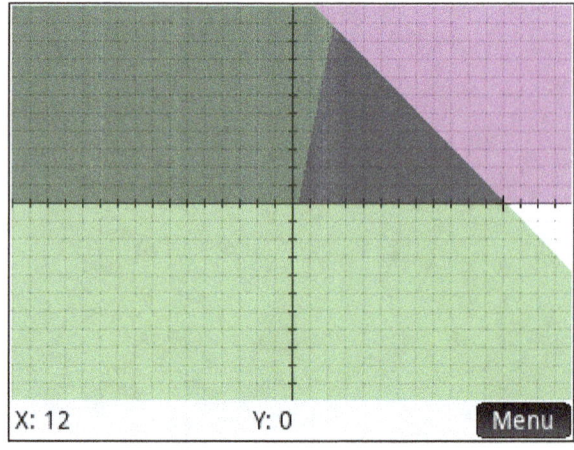

| $x + y \leq 12$ | $y \geq 0$ | (12, 0) |

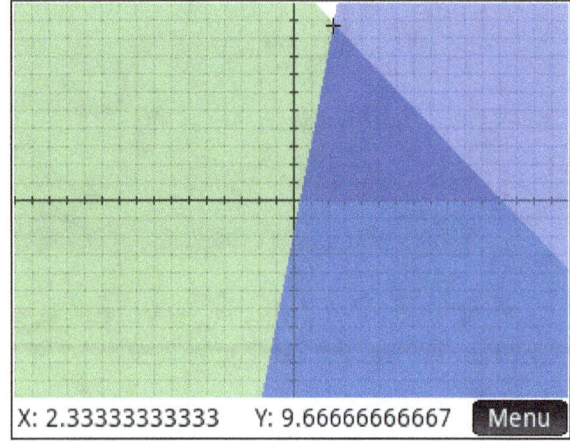

| $5x - y \geq 2$ | $5x - y \geq 2$ | (7/3, 29/3) |

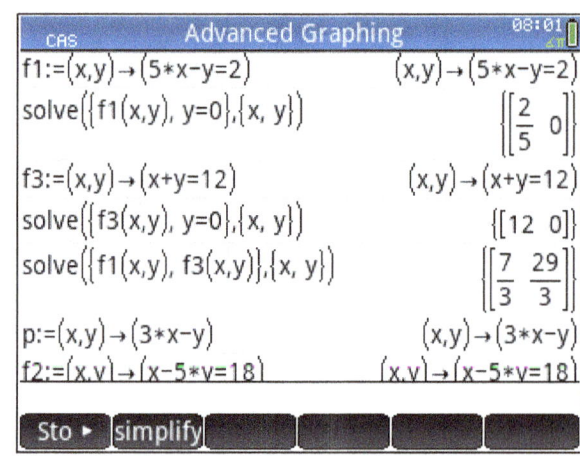

| Corners | (2/5, 0) | (12, 0) | (7/3, 29/3) |

Graphical Minimums

Thursday, May 16, 2019   2:58 PM

http://computerlearningservice.com/Academy/Finite-Math/Dual-Problem/Graphical-Minimums/graphical-minimums.html

Unbounded - Standard

Thursday, May 16, 2019   4:46 AM

Technical Specifications - Daily Diet - Unit (ounce)

	Food$_1$	Food$_2$	Requirements
Vitamin A per unit	2	4	22
Vitamin B per unit	3	2	20
Iron per unit	4	5	40
Cost per unit	6 cents	5 cents	

Minimize $C = 6x + 5y$

Subject to $x \geq 0, y \geq 0$

$2x + 4y \geq 22$

$3x + 2y \geq 20$

$4x + 5y \geq 40$

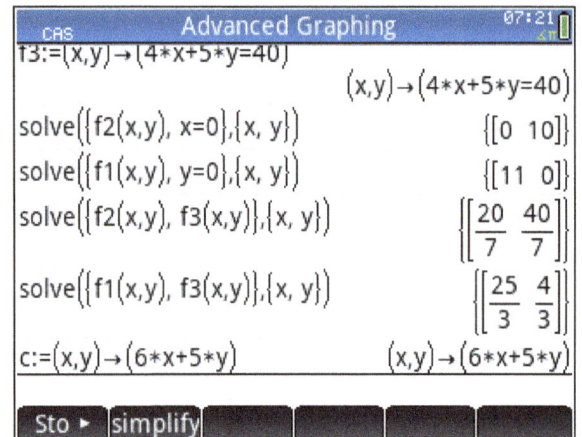
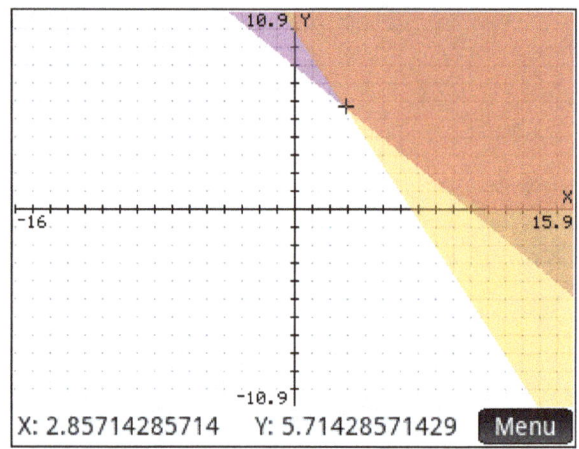

Corners	(0, 10)	(11, 0)	$\left(\dfrac{20}{7}, \dfrac{40}{7}\right)$	$\left(\dfrac{25}{3}, \dfrac{4}{3}\right)$
	50	66	$45\dfrac{5}{7}$	$56\dfrac{2}{3}$

The minimum occurs at (20/7, 40/7) with a value of 45 5/7. Purchase Food$_1$, approximately 2.86 ounces, and Food$_2$, approximately 5.71 ounces, to meet daily requirements at a cost of approximately 46 cents.

Subpage - Intercepts

Thursday, May 16, 2019   5:24 AM

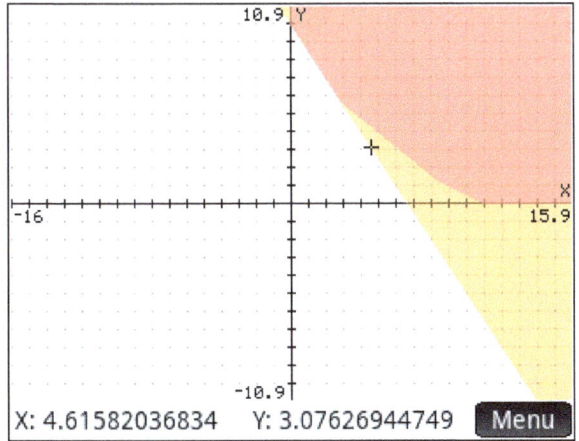

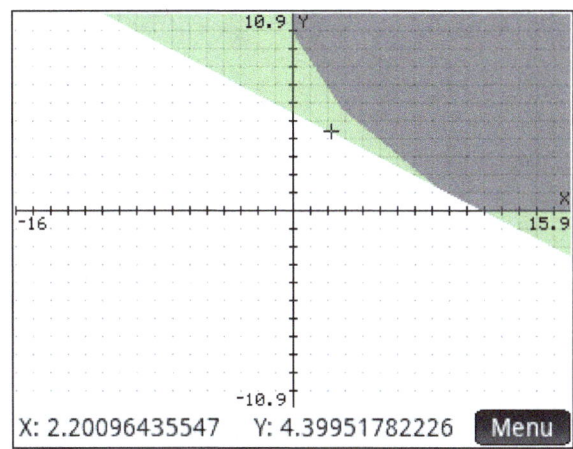

$3x + 2y \geq 20$, (0,10)

$2x + 4y \geq 22$, (11,0)

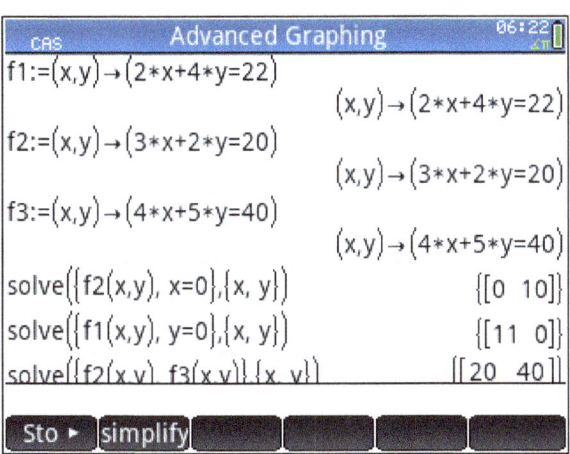

| Corners | (0,10) | (11,0) |

Subpage - Simultaneous Equations

Thursday, May 16, 2019   6:21 AM

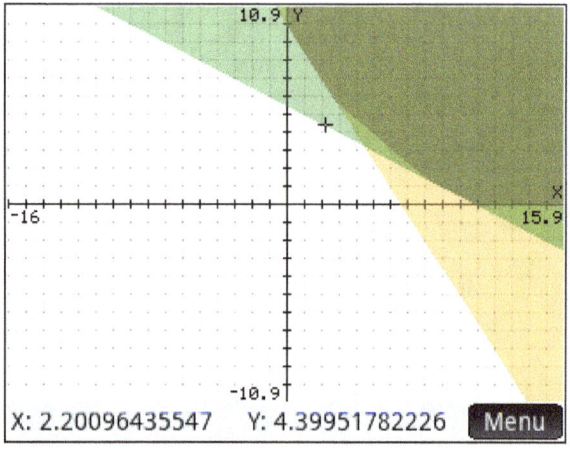

V3 and V4 Outside region

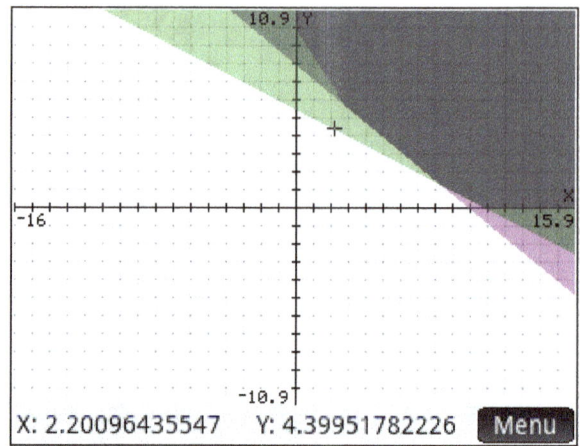

V3 and V5 $\left(\dfrac{20}{7}, \dfrac{40}{7}\right)$

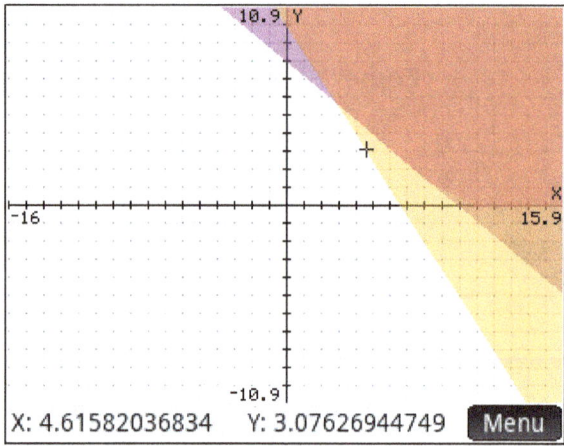

V4 and V5 $\left(\dfrac{25}{3}, \dfrac{4}{3}\right)$

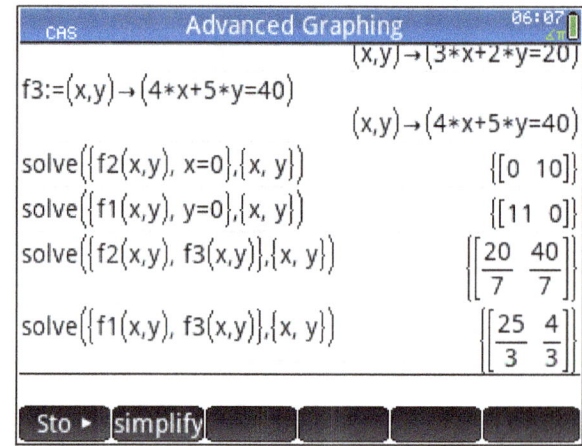

| Corners | $\left(\dfrac{20}{7}, \dfrac{40}{7}\right)$ | $\left(\dfrac{25}{3}, \dfrac{4}{3}\right)$ |

Linear Programing Page 63

Line - Nonstandard

Tuesday, May 21, 2019   9:57 AM

Minimize $C = 3x + 2y$

Subject to $x \geq 0, y \geq 0$

$x + y \leq 5$

$3x + 7y \geq 21$

$-x + y = 2$

Maximize $P = -C = -3x - 2y$

Subject to $x \geq 0, y \geq 0$

$x + y \leq 5$

$-3x - 7y \leq -21$

$-x + y \geq 2$ $x - y \leq -2$

$-x + y \leq 2$

Minimum of 15/2 at x =7/10 and y =27/10.

X	Y	U	V	W	T	P	Constraints
1	1	1	0	0	0	0	5
-3	-7	0	1	0	0	0	21
1	-1	0	0	1	0	0	-2
-1	1	0	0	0	1	0	2
3	2	0	0	0	0	1	0

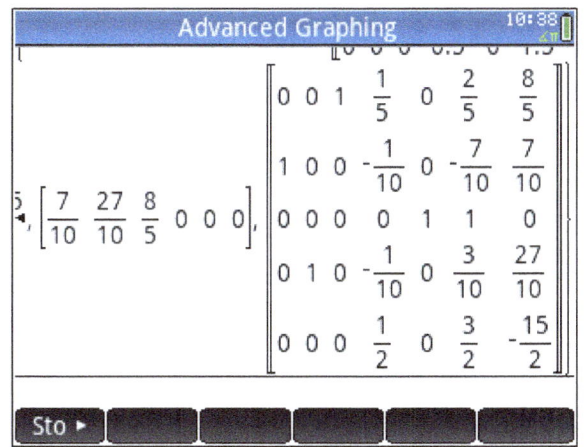
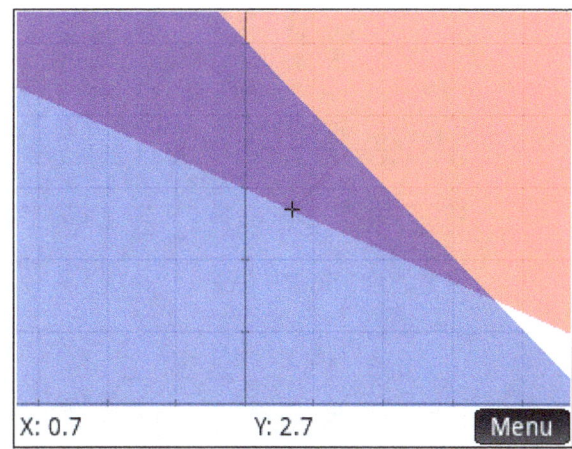

Subpage

Tuesday, May 21, 2019   10:59 AM

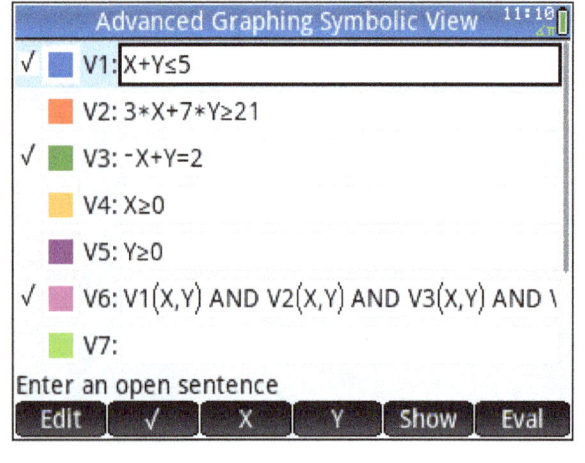
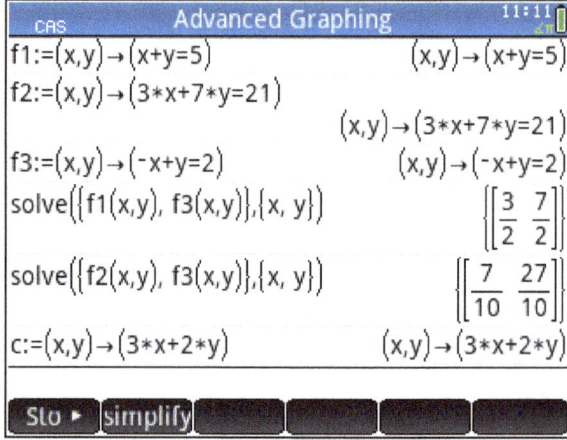
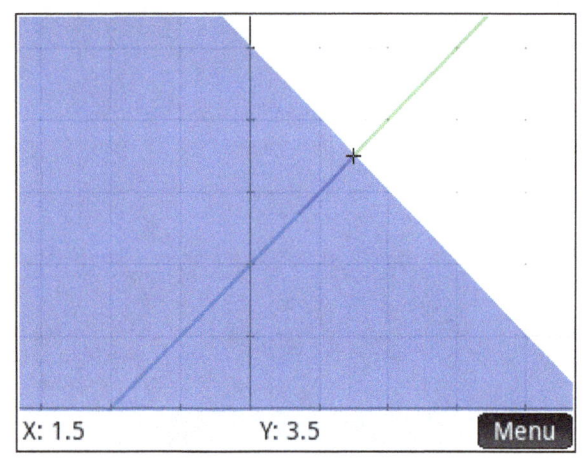
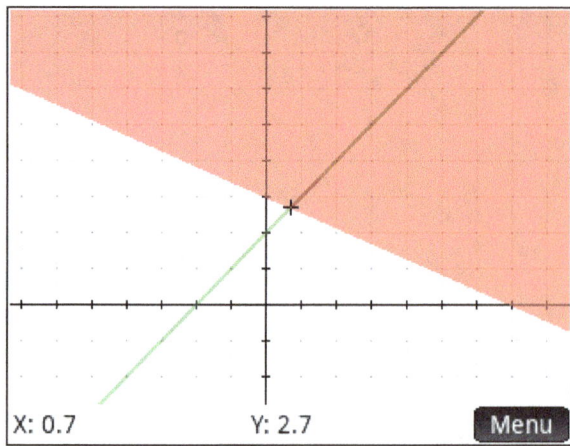

| $x+y \leq 5$ | $-x+y=2$ | (3/2, 7/2) |

| $-3x - 7y \leq 21$ | $-x+y=2$ | (7/10), 27/10) |

End Points	(3/2, 7/2)	(7/10), 27/10)
c(x,y)	23/2	15/2

Minimum of 15/2 at x = 7/10 and y = 27/10.

Simplex Maximums

Wednesday, May 15, 2019   12:27 AM

http://computerlearningservice.com/Academy/Finite-Math/Dual-Problem/Simplex-Maximums/simplex-maximums.html

Standard

Wednesday, May 15, 2019   12:28 AM

To satisfy stockholders a company needs to make a profit of $5 on each goose down pillow and $9 on each goose down comforter it manufactures. The company buys high-quality goose down from a distributor who guarantee a supply of 2,500 pounds a week. The factory is limited to 45 workers at 40 hours per week. The pillow requires 2 pounds of down, 1 hour of work and comforter 4 pounds of down, 3 hours of work. How many of each per week should the company manufacture for maximum profit?

	Per pillow	Per comforter	Total
Down	2 pounds	4 pounds	2500 pounds
Time	1 hours	3 hours	1800 hours
Profit	$5	$9	

Maximize $P = 5x + 9y$

Subject to $x \geq 0, y \geq 0$

$2x + 4y \leq 2500$

$x + 3y \leq 1800$

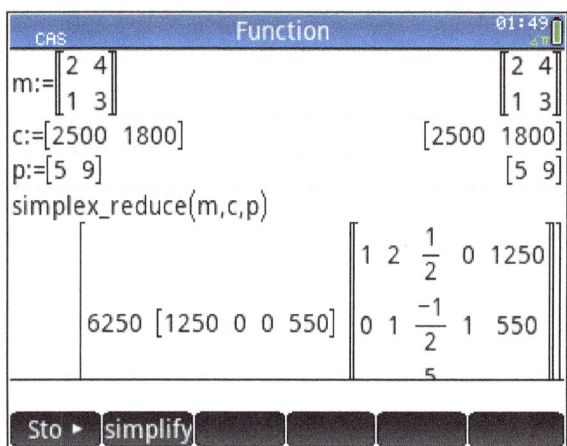

The company should make 1,250 goose down pillows per week for a profit of $6,250.

Subpage

Wednesday, May 15, 2019   2:07 AM

Maximize $P = 5x + 9y$
Subject to $x \geq 0, y \geq 0$
$2x + 4y \leq 2500$
$x + 3y \leq 1800$

The company should make 1,250 goose down pillows per week for a profit of $6,250.

X	Y	U	V	P	Constraints
2	4	1	0	0	2500
1	3	0	1	0	1800
-5	-9	0	0	1	0

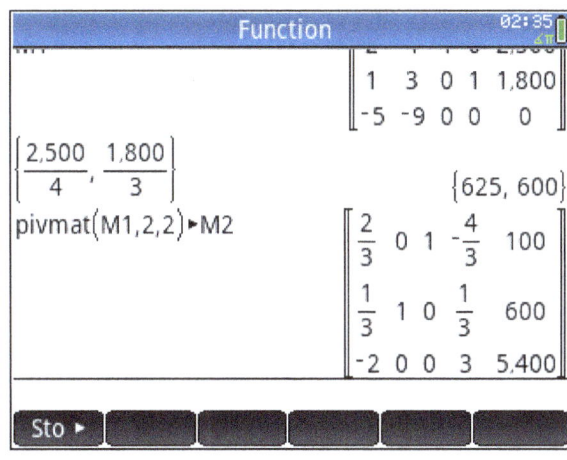

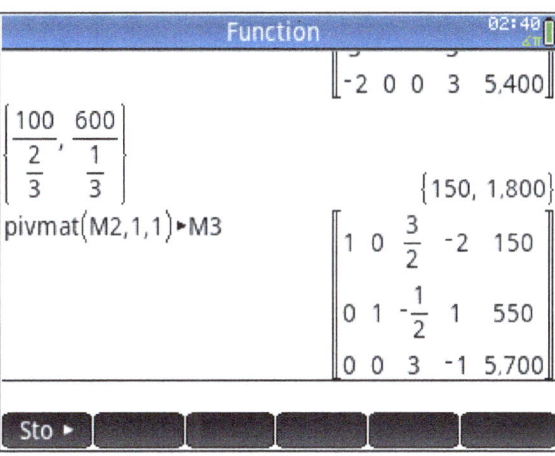

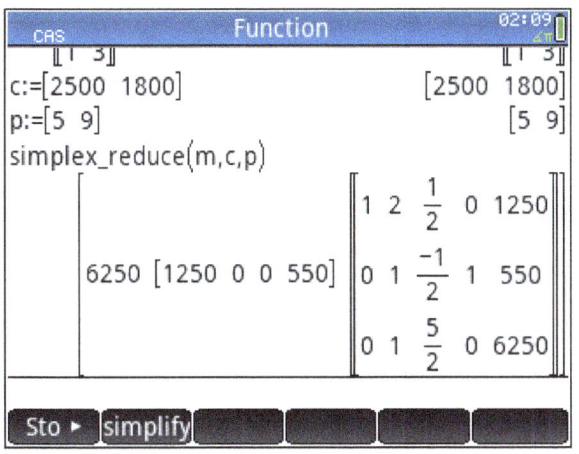

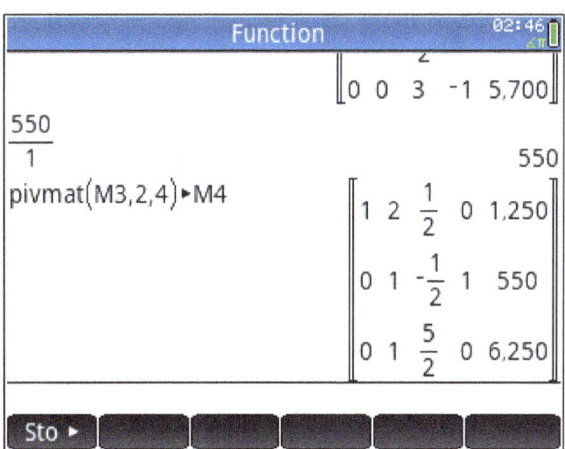

Nonstandard

Monday, May 20, 2019   11:01 PM

Maximize $P = 3x - y$
Subject to $x \geq 0, y \geq 0$
$5x - y \geq 2$
$x - 5y \leq 18$
$x + y \leq 12$

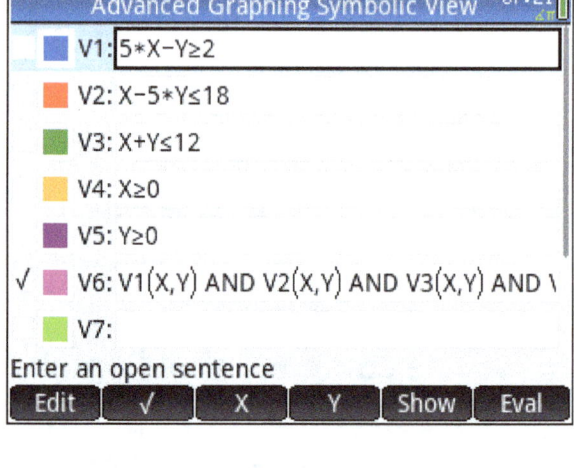

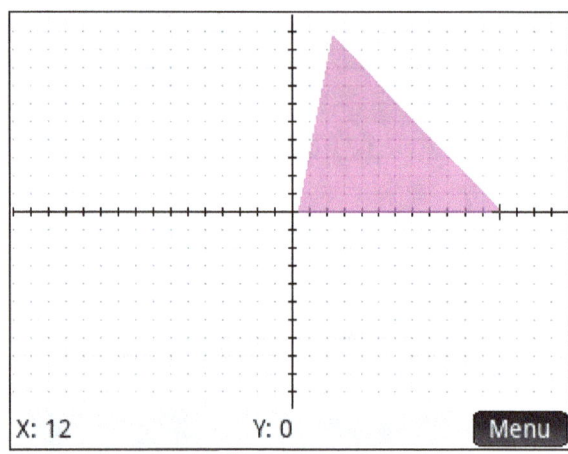

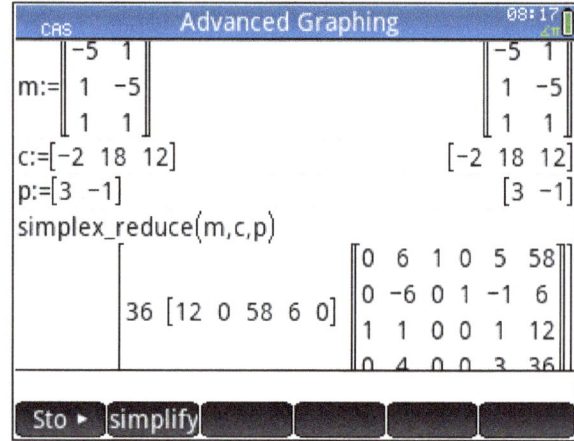

Corners	(2/5, 0)	(12, 0)	(7/3, 29/3)
p(x,y)	6/5	36	-8/3

Maximum of 36 at x=12 and y=0. Note simplex_reduce correct answer though it is not applicable.

simplex_reduce(Matrix_A, Vector_B, Vector_C) is not applicable. $5x - y \geq 2 \rightarrow -5x + y \leq -2$

Simplex Reduction help specifies: A_x <= 0 and x>=0, b>= 0. c(1,1)=-2 violates b>=0.

Subpage

Monday, May 20, 2019   11:10 PM

Maximize $P = 3x - y$

Subject to $x \geq 0, y \geq 0$

$-5x + y \leq -2$

$x - 5y \leq 18$

$x + y \leq 12$

X	Y	U	V	W	P	Constraints
-5	1	1	0	0	0	-2
1	-5	0	1	0	0	18
1	1	0	0	1	0	12
-3	1	0	0	0	1	0

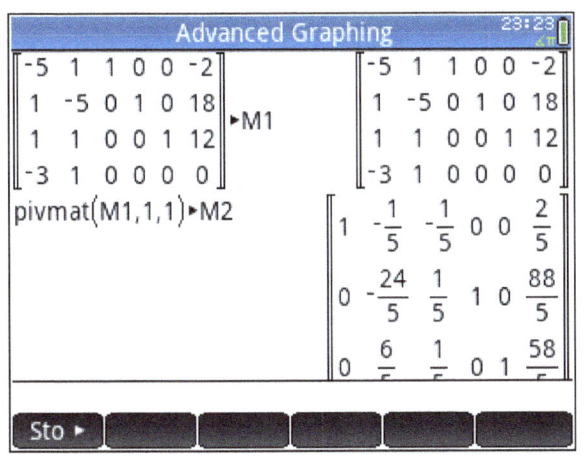

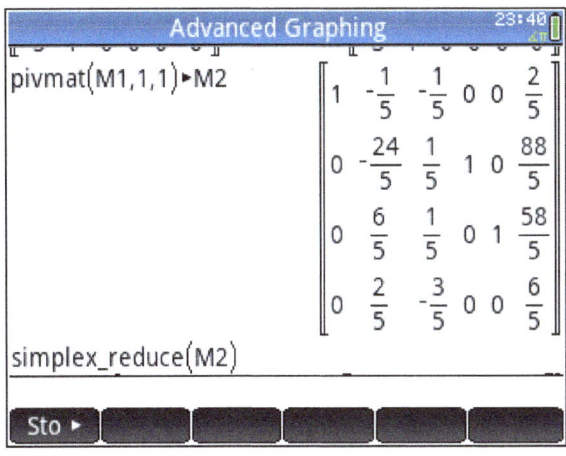

Corners	(0, 0)	(2/5, 0)	(12, 0)
	0	6/5	36
Matrix	M1	M2	M3
Feasible	no	yes	yes

Maximum of 36 at x=12 and y=0. Note simplex_reduce correct answer same as manual pivmat.

Linear Programing Page 70

Simplex Minimums

Thursday, May 16, 2019   3:10 PM

http://computerlearningservice.com/Academy/Finite-Math/Dual-Problem/Simplex-Minimums/simplex-minimums.html

The Dual

Thursday, May 16, 2019   3:11 PM

Technical Specifications - Daily Diet - Unit (ounce)

	Food$_1$	Food$_2$	Requirements
Vitamin A per unit	2	4	22
Vitamin B per unit	3	2	20
Iron per unit	4	5	40
Cost per unit	6 cents	5 cents	

Minimize $C = 6x + 5y$

Subject to $x \geq 0, y \geq 0$

$2x + 4y \geq 22$

$3x + 2y \geq 20$

$4x + 5y \geq 40$

X	Y	Constant
2	4	22
3	2	20
4	5	40
6	5	

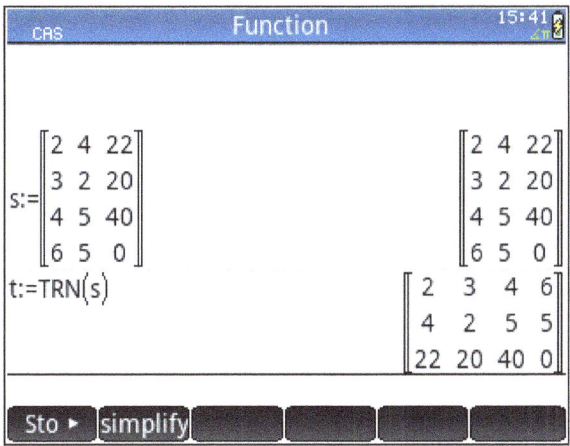

Maximize $P = 22u + 20v + 40w$

Subject to $u \geq 0, v \geq 0, w \geq 0$

$2u + 3v + 4w \leq 6$

$4u + 2v + 5w \leq 5$

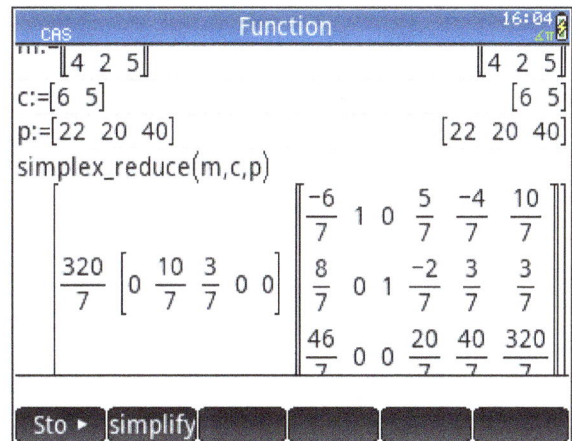

The minimum occurs at (20/7, 40/7) with a value of 45 5/7. Purchase Food$_1$, approximately 2.86 ounces, and Food$_2$, approximately 5.71 ounces, to meet daily requirements at a cost of approximately 46 cents.

Subpage

Thursday, May 16, 2019   4:09 PM

Maximize $P = 22u + 20v + 40w$

Subject to $u \geq 0, v \geq 0, w \geq 0$

$2u + 3v + 4w \leq 6$

$4u + 2v + 5w \leq 5$

The minimum occurs at (20/7. 40/7) with a value of 45 5/7. Purchase Food$_1$, approximately 2.86 ounces, and Food$_2$, approximately 5.71 ounces, to meet daily requirements at a cost of approximately 46 cents.

U	V	W	X	Y	p	Constraint
2	3	4	1	0	0	6
4	2	5	0	1	0	5
-22	-20	-40	0	0	1	0

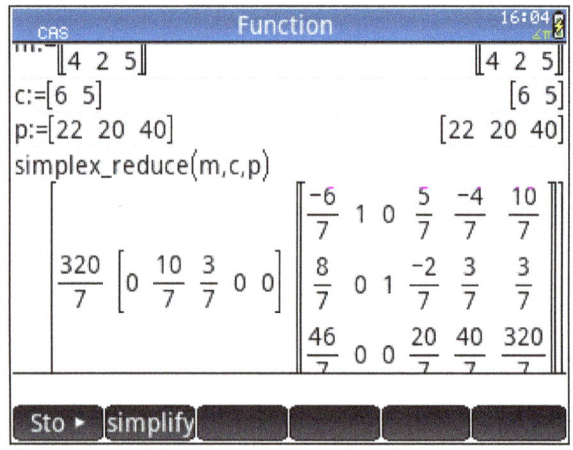

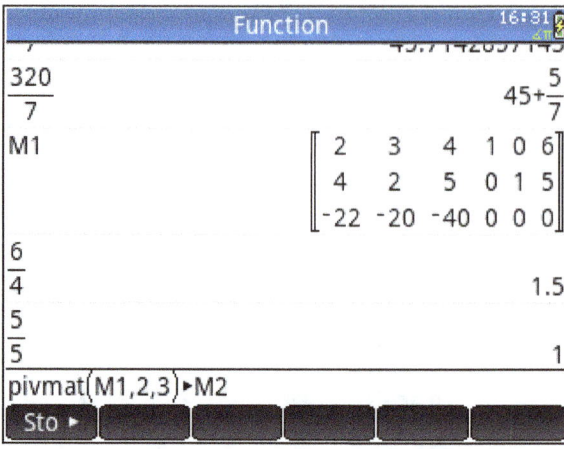

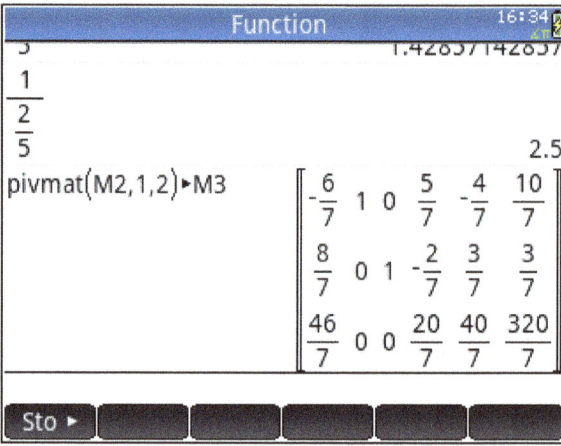

Line - Nonstandard

Tuesday, May 21, 2019   9:57 AM

Minimize $C = 3x + 2y$
Subject to $x \geq 0, y \geq 0$
$x + y \leq 5$
$3x + 7y \geq 21$
$-x + y = 2$

Maximize $P = -C = -3x - 2y$
Subject to $x \geq 0, y \geq 0$
$x + y \leq 5$
$-3x - 7y \leq -21$
$-x + y \geq 2$ $x - y \leq -2$
$-x + y \leq 2$

Minimum of 15/2 at x =7/10 and y =27/10.

X	Y	U	V	W	T	P	Constraints
1	1	1	0	0	0	0	5
-3	-7	0	1	0	0	0	21
1	-1	0	0	1	0	0	-2
-1	1	0	0	0	1	0	2
3	2	0	0	0	0	1	0

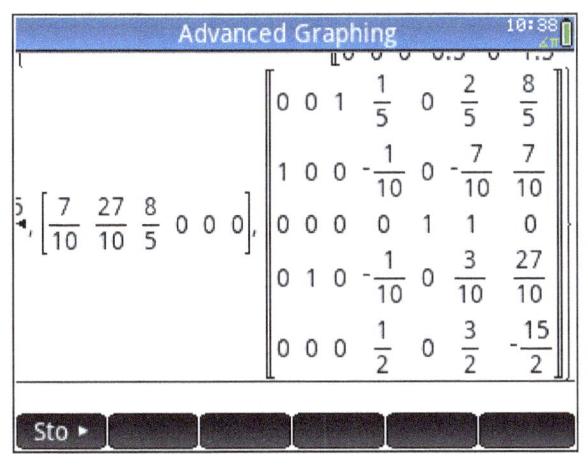
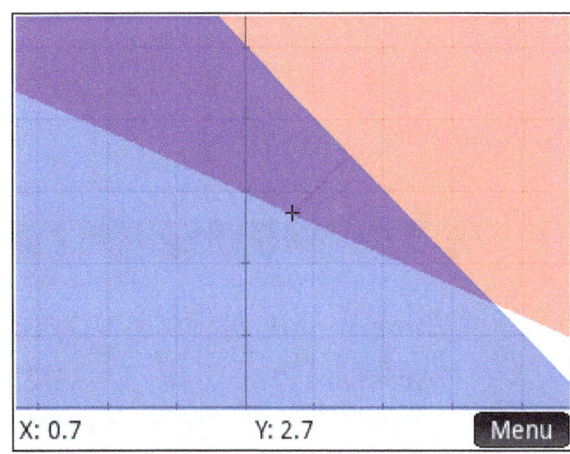

Subpage

Tuesday, May 21, 2019   10:47 AM

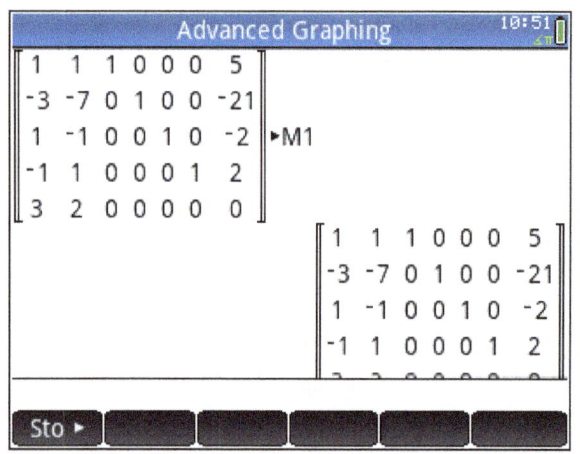
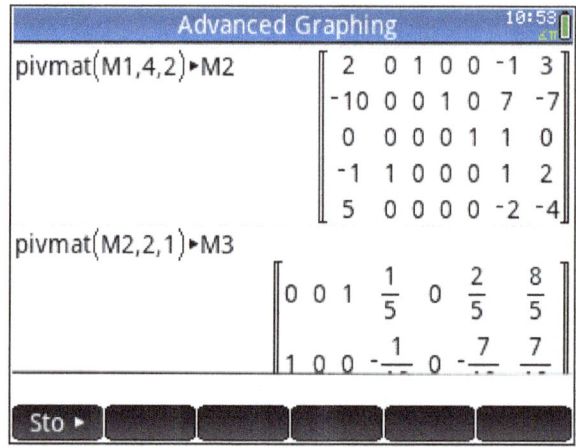
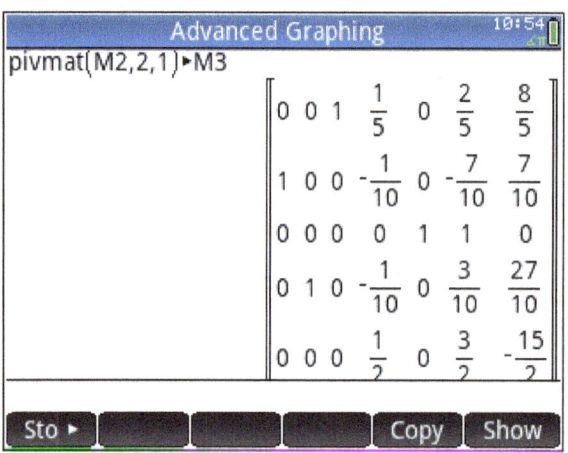
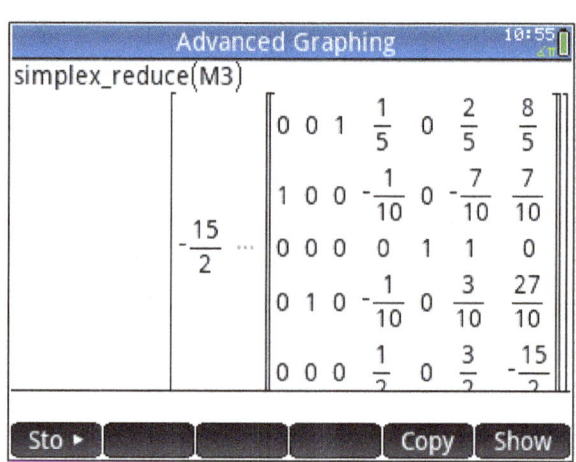

Corners	(0, 0)	(0, 2)	(7/10), 27/10)
Maximums	0	-4	-15/2
Matrix	M1	M2	M3
Feasible	no	no	yes

$C = -P$ Minimum of 15/2 at x = 7/10 and y = 27/10.

5

Sets and Probabilities

"If there a 50-50 chance that
something will go wrong then
nine times out of 10 it will."
~ Paul Harvey

Sets

- Set Operations 77
- Orientations 78
- Set Algebra 80
- De Morgan's Law 81

Number Elements Finite Set

- Count for A and B 83
- Applications 84

Multiplication Principle

- Coin Tosses 87
- Applications 88

Permutations and Combinations

- Introduction   90
- Permutations   91
- Combinations   92
- Explorations   93

Experiment or Observation

- Terminology   95
- Venn Diagrams   97

Probability

- Theoretical and Subjective   99
- Additional Topics   100

Rules of Probability

- Basic Properties
- Computations   101

Sets

Saturday, June 22, 2019   9:17 AM

http://computerlearningservice.com/Academy/Finite-Math/Sets-Probability/Sets/sets.html

Set Operations

Friday, June 21, 2019   7:23 AM

Universal Set - Union L1,L2,L3

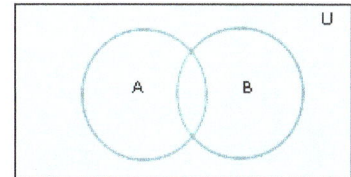

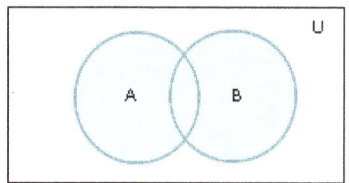
Union L1, L2

Proper Subset A   Intersection L1,l2   Complement B

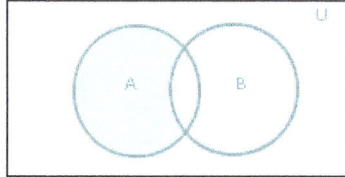 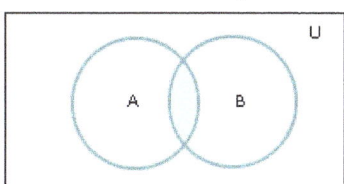 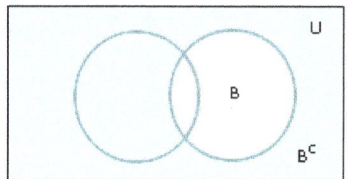

L1, L2, l3

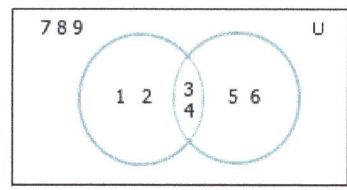

Diagram	HP Prime	Set Notation
A	L1	{1, 2, 3, 4}
B	L2	{3, 4, 5, 6}
$(A \cup B)^c$	L3	{7, 8, 9}
U	UNION(L1,L2,L3) ▷ u	{1, 2, 3, 4, 5, 6, 7, 8, 9}
$A \cup B$	UNION(L1,L2) ▷ L4	{1. 2. 3, 4, 5, 5}
$(A \cap B^c) \subset (A \cup B)$	prsubL1(L2,L4) ▷ L5	{1, 2}
$A \cap B$	INTERSECT(l1,L2,L3) ▷ L6	{3, 4}
B^c	Complement(u,L2)	{1, 2, 7, 8, 9}

prsubL1(a,b):=DIFFERENCE(a,b)

prsubL2(a,b):=DIFFERENCE(a,b)

Complement(a,b):=DIFFERENCE(a,b)

Orientations

Saturday, June 22, 2019   3:09 AM

$A = B$

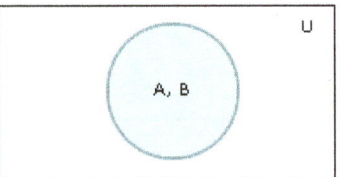

Diagram	HP Prime	Set Notation
A	L1	{1, 2, 3, 4}
B	L2	{1, 2, 3, 4}
A^c	L3	{5, 6, 7, 8, 9}

$B \subset A$

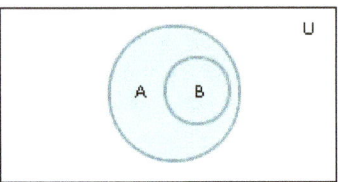

Diagram	HP Prime	Set Notation
A	L1	{1, 2, 3, 4}
B	L2	{3, 4}
A^c	L3	{5, 6, 7, 8, 9}

$A \cap B = \emptyset$

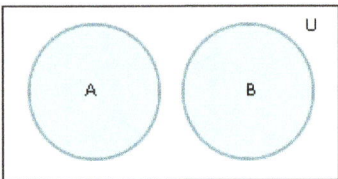

Diagram	HP Prime	Set Notation
A	L1	{1, 2, 3}
B	L2	{4, 5, 6}
$(A \cup B)^c$	L3	{7, 8, 9}

$A \cap B \neq \emptyset$

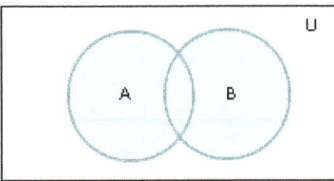

Diagram	HP Prime	Set Notation
A	L1	{1, 2, 3, 4}
B	L2	{3, 4, 5, 6}
$(A \cup B)^c$	L3	{7, 8, 9}

$\emptyset \subset A$ Proper Subset
$B \subseteq A$ Subset
$B \subset A$ Proper Subset
$A \subseteq A$ Subset

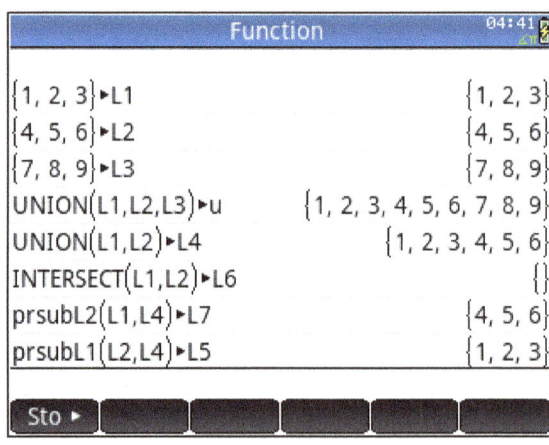

Sets and Probabilities Page 78

Subpage

Wednesday, April 27, 2022   5:10 AM

Continuation of previous page

```
{1, 2, 3, 4}▶L1                    {1, 2, 3, 4}
{3, 4, 5, 6}▶L2                    {3, 4, 5, 6}
{7, 8, 9}▶L3                          {7, 8, 9}
UNION(L1,L2,L3)▶u     {1, 2, 3, 4, 5, 6, 7, 8, 9}
UNION(L1,L2)▶L4              {1, 2, 3, 4, 5, 6}
INTERSECT(L1,L2)▶L6                     {3, 4}
prsubL2(L1,L4)▶L7                       {5, 6}
```

Set Algebra

Saturday, June 22, 2019   8:38 AM

Complements

$\emptyset^c = U$	$U^c = \emptyset$	$(A^c)^c = A$
$A \cap A^c = \emptyset$	$A \cup A^c = U$	

Operations

$A \cap B = B \cap A$	Commutative Intersection
$A \cup B = B \cup A$	Commutative Union
$(A \cap B) \cap C = A \cap (B \cap C)$	Associate Intersection
$(A \cup B) \cup C = A \cup (B \cup C)$	Associate Union
$A \cap (B \cup C)$ $= (A \cap B) \cup (A \cap C)$	Distributive Intersection
$A \cup (B \cap C)$ $= (A \cup B) \cap (A \cup C)$	Distributive Union

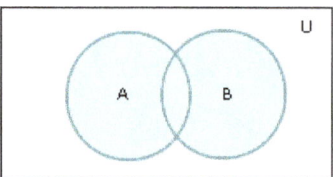

Diagram	HP Prime	Set Notation
A	L1	{1, 2, 3, 4}
B	L2	{3, 4, 5, 6}

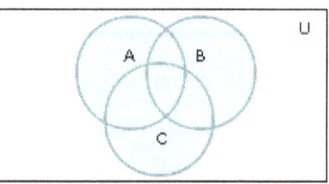

Diagram	HP Prime	Set Notation
A	L1	{1, 2, 3, 4}
B	L2	{2, 4, 5, 6}
C	L3	{3, 4, 6, 7}

Sets and Probabilities Page 80

De Morgan's Law

Saturday, June 22, 2019   11:13 PM

De Morgan's Law	
$(A \cap B)^c = A^c \cup B^c$	Intersection
$(A \cup B)^c = A^c \cap B^c$	Union

$(A \cap B)^c$ - L6

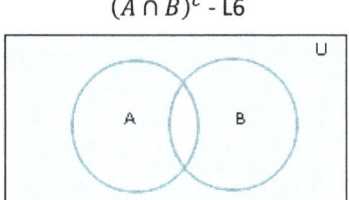

A^c - L7 B^c - L8

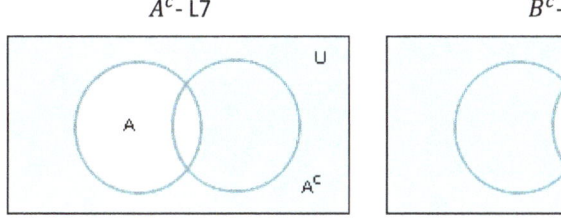

Join Left and Right Drawings - L9

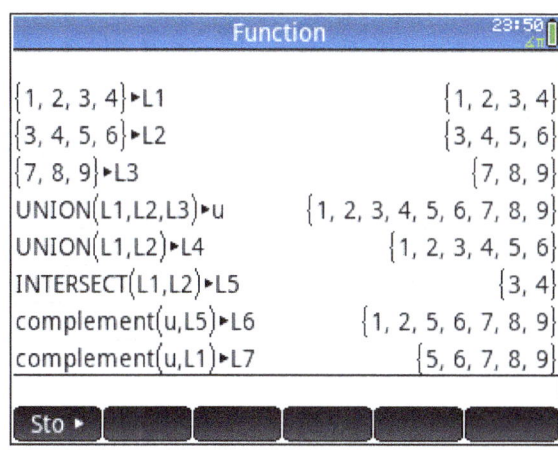

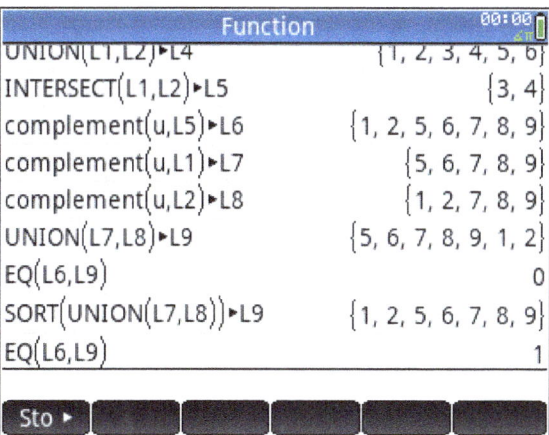

Sets and Probabilities Page 81

Number Elements Finite Set

Tuesday, June 25, 2019   11:13 PM

http://computerlearningservice.com/Academy/Finite-Math/Sets-Probability/Number-Elements/number-elements.html

Count for A and B

Saturday, June 22, 2019   3:09 AM   Aving

$n(A \cup B) = n(A) + n(B)$

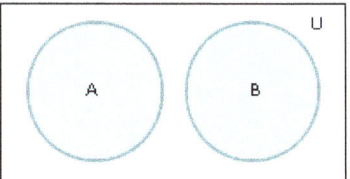

Diagram	HP Prime	Set Notation
A	L1	{1, 2, 3, 4}
B	L2	{5, 6, 7, 8}
$(A \cup B)^c$	L3	{9}

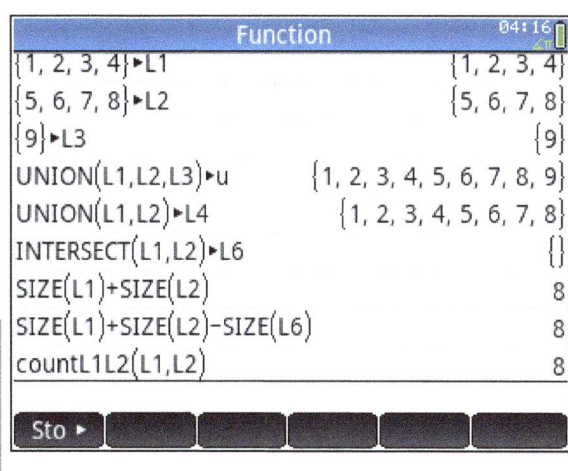

$n(A \cup B) = n(A) + n(B) - n(A \cap B)$

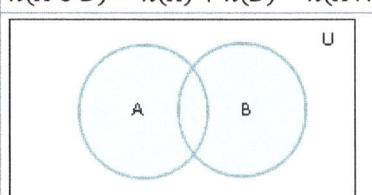

Diagram	HP Prime	Set Notation
A	L1	{1, 2, 3, 4}
B	L2	{3, 4, 5, 6}
$(A \cup B)^c$	L3	{7, 8, 9}

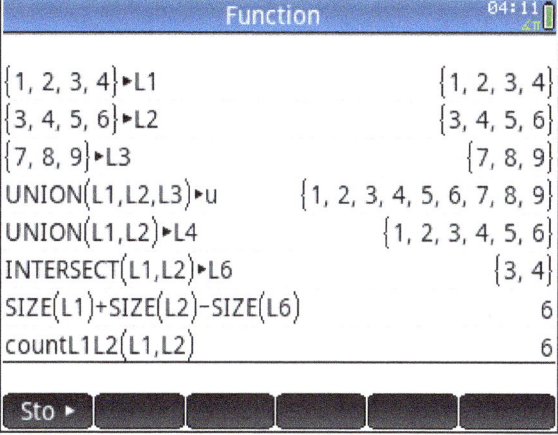

countL1L2(a,b):=size(a)+size(b)-size(INTERSECT(a,b))

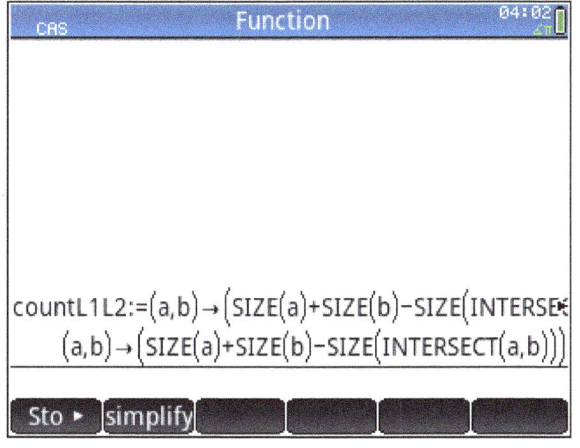

Sets and Probabilities Page 83

Applications

Wednesday, June 26, 2019   4:45 AM

1. A survey of 1002 adults nationwide in 2018 produced the following information: 625 higher priority to reduce gun violence, 395 higher priority to own guns, and 50 favored both. How many favored reduce gun violence but not protect gun ownership?

$n(A) - n(A \cap B) = 625 - 50$

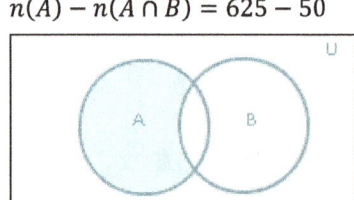

Diagram		Count
A	Reduce violence	625
B	Own guns	395
$A \cap B$	Both	50
$(A \cup B)^c$	Unsure or neither	32

We are looking at the intersection of A with B complement. From the Venn diagram the answer is 575.

2. A mess hall conducted a survey to help with food and supplies needed. The following data was obtained:

   55 soldiers ate all three meals; 115 ate breakfast and dinner; 88 ate lunch and dinner; 77 ate breakfast and lunch; 140 ate breakfast; 182 ate lunch; 272 ate dinner

   How many soldiers:
   a. Ate only dinner at the mess hall?
   b. Ate exactly one meal at the mess hall?
   c. Ate exactly two meals in the mess hall?
   d. Ate at least one meal in the mess hall

Subpage

Thursday, June 27, 2019   3:52 AM

All three meals

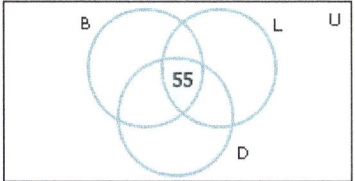

Exactly two meals

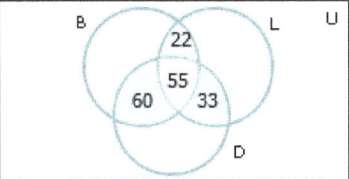

Exactly one meal

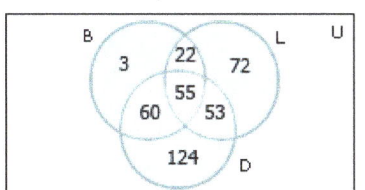

Two or more meals

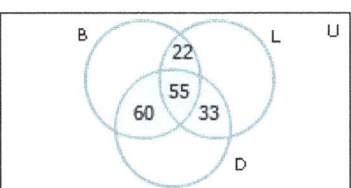

At least one meal

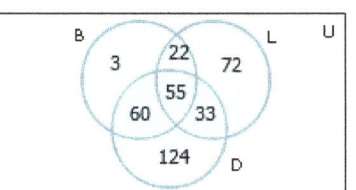

$n(B) = 140 \qquad n(B \cap D) = 60 + 55 = 115$

$n(L) = 182 \qquad n(L \cap D) = 33 + 55 = 88$

$n(D) = 272 \qquad n(B \cap L) = 22 + 55 = 77$

$n(B \cap L \cap D) = 55$

$n(B \cap D \cap L^c) = 115 - 55 = 60$

$n(L \cap D \cap B^c) = 88 - 55 = 33$

$n(B \cap L \cap D^c) = 77 - 55 = 22$

$n(B \cap L^c \cap D^c) = 140 - (22 + 55 + 60) = 3$

$n(D \cap B^c \cap L^c) = 272 - (60 + 55 + 33) = 124$

$n(L \cap D^c \cap B^c) = 182 - (22 + 55 + 33) = 72$

How many soldiers:

a. Ate only dinner at the mess hall? 124
b. Ate exactly one meal at the mess hall? 3+72+124=199
c. Ate exactly two meals in the mess hall? 22+60+33=115
d. Ate at least one meal in the mess hall? 55+115+199=369

Sets and Probabilities Page 85

Multiplication Principle

Tuesday, July 2, 2019   4:03 AM

http://computerlearningservice.com/Academy/Finite-Math/Sets-Probability/Multiplication-Principle/multiplication-principle.html

Coin Tosses

Tuesday, July 2, 2019   4:18 AM

Multiplication Principle

Suppose there are m possible outcomes for task 1 and n possible outcome for task 2, then are $m \times n$ outcomes for task 1 followed by task 2. Principle can be extended to a finite number of tasks.

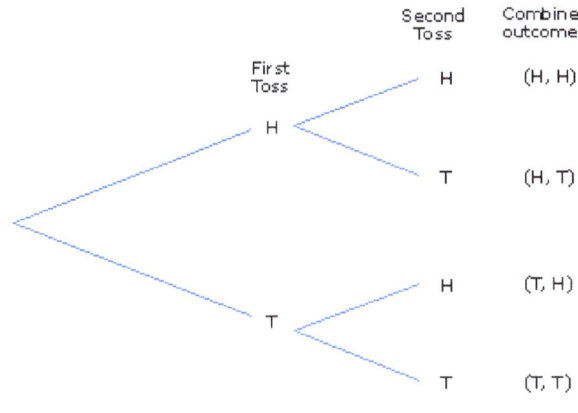

$2 \times 2 = 4$

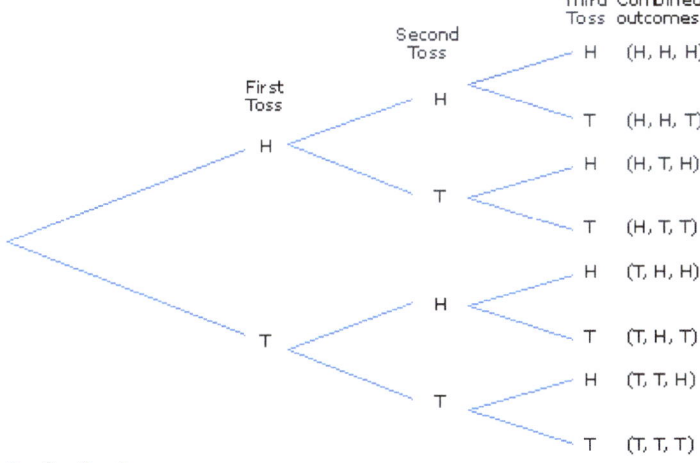

$2 \times 2 \times 2 = 8$

Applications

Tuesday, July 2, 2019   3:21 PM

1. A restaurant has a summer menu containing three choices for appetizer, five main entrees, and four desserts. How many special summer meals are possible?

   $3 \times 5 \times 4 = 60$

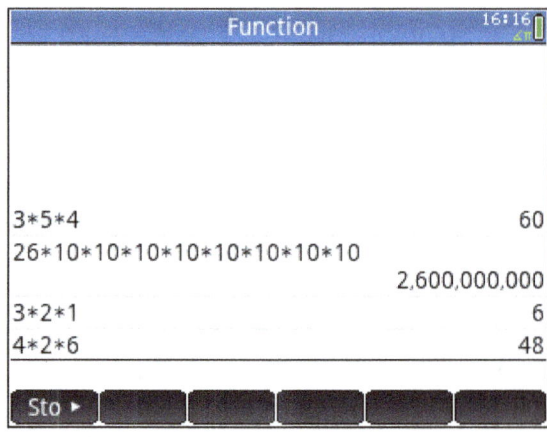

2. The military in Canada uses a Service Number that is the same format as the old Social Insurance Number (SIN) so as to be able to keep using same service forms. The new Service number's form is X12 345 678, with X representing a random alphabetic letter followed by 8 numbers. How many Service Numbers are possible?

   $26 \times 10 \times 10 \times 10 \times 10 \times 10 \times 10 \times 10 \times 10 =$
   $2{,}600{,}000{,}000$

3. For a show of four local artist the city's Civic Art center wishes to display five painting by these four artist on a wall. The two painting by the same artist are to be displayed next to each other. In how many ways can the paintings be hung?

   Task 1: There are four adjacent spots for the artist with two painting. One and two, two and three, three and four, and four and five. Task 2: There are two ways to hang his painting for each adjacent spot. Task 3: The remaining three paintings hung in three remaining spots, $3 \times 2 \times 1 = 6$.
   $4 \times 2 \times 6 = 48$

Permutations and Combinations

Wednesday, July 3, 2019   5:27 AM

http://computerlearningservice.com/Academy/Finite-Math/Sets-Probability/Permutations-and-Combinations/permutations-and-combinations.html

Introduction

Wednesday, July 3, 2019   5:28 AM

Factorials

$1! = 1$

$2! = 2 \cdot 1 = 2$

$3! = 3 \cdot 2 \cdot 1 = 6$

$n! = n(n-1)(n-2) \cdots 3 \cdot 2 \cdot 1$

So that certain formulas hold we define $0! = 1$.

How many ways can letters *a*, *b*, and *c* be arranged?

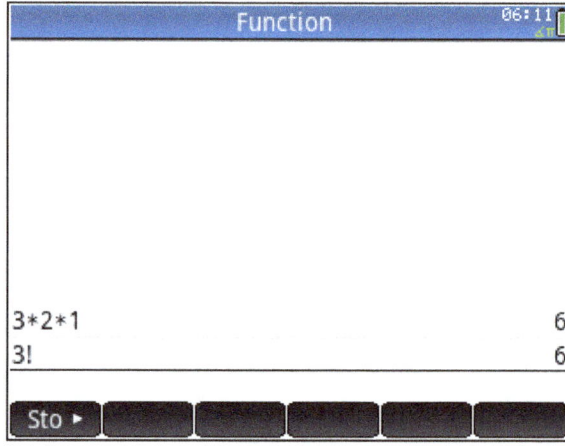

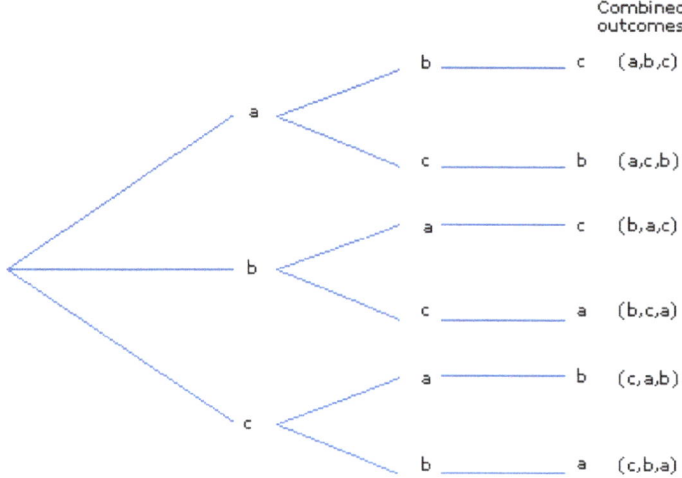

List them: $abc, acb, bac, bca, cab, cba$
Six arrangements for the three letters.

Use multiplication principle: $3 \cdot 2 \cdot 1 = 6$

Permutations

Wednesday, July 3, 2019   6:15 AM

Permutations

We have permutations when the selection comes from a single group without repetition with the order of the elements is taken into account. Example: abc is different than cba.

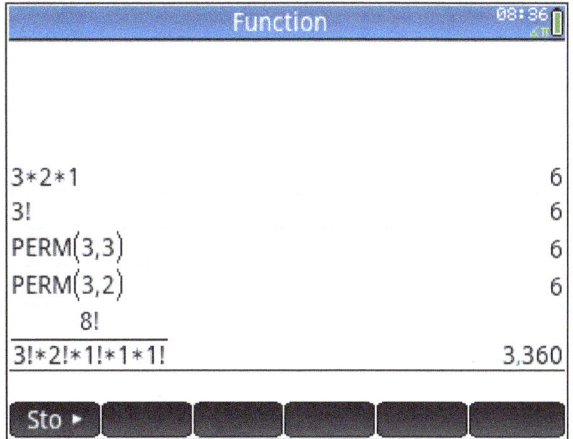

If all elements of the group are selected in the permutation, we have:
$P(n,n) = n!$

Permutation Formula

$$P(n,r) = \frac{n!}{(n-r)!}$$

The number of n things taken r at a time.

Two letters selected from our original group of three.

$$P(3,2) = \frac{3!}{(3-2)!} = \frac{3!}{1!} = 3! = 3 \cdot 2 \cdot 1 = 6$$

List them: ab, ac, ba, bc, ca, cb
Six arrangements for the two letters.

Non-distinguishable Objects

$r_1 + r_2 + \cdots + r_k = n$

$$\frac{n!}{r_1! r_2! \cdots r_k!}$$

How many different permutations of the letters in word ILLINOIS?

$$\frac{8!}{3!\, 2!\, 1!\, 1!\, 1!} = 3,360$$

Combinations

Wednesday, July 3, 2019   8:00 AM

Combinations
Combinations occurs when the selection comes from a single group without repetition with the order of the elements is not taken into account. Example: cba is considered the same as abc.

Combination Formula
$$C(n,r) = \frac{n!}{r!(n-r)!}$$

The number of n things taken r at a time.

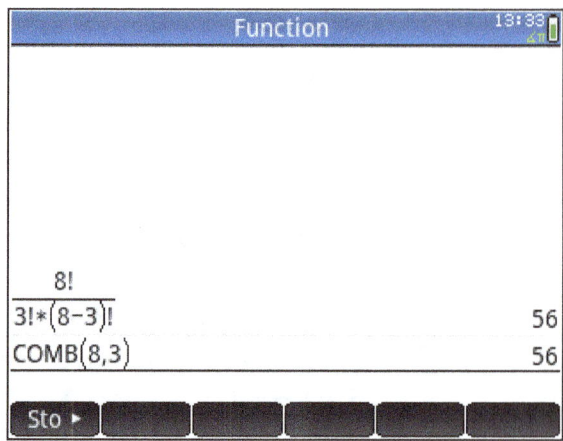

How many different three people committees be chosen form eight people?

$$C(8,3) = \frac{8!}{3!(8-3)!} = \frac{8!}{3!\,5!} = \frac{8 \cdot 7 \cdot 6}{3 \cdot 2 \cdot 1} = 8 \cdot 7 = 56$$

For large values we use the formula. Listing the combinations is not useful.

Remember if order is important, permutations should be used, but if order is not important, combinations should be used.

Explorations

Permutation Formula
$$P(n,r) = {}_nP_r = \frac{n!}{(n-r)!}$$
The number of n things taken r at a time.

Non-distinguishable Objects
$$r_1 + r_2 + \cdots + r_k = n$$
$$\frac{n!}{r_1! r_2! \cdots r_k!}$$

Combination Formula
$$C(n,r) = {}_nC_r = \frac{n!}{r!(n-r)!}$$
The number of n things taken r at a time.

Evaluate

$P(6,3) = 120 \qquad C(8,4) = 70$

21st Century Auto has received 6 inquires about the new Model Z electric car. How many ways can the inquires be directed to three of the dealer's salesman if each salesman is to handle two inquires?

$\dfrac{6!}{2!2!2!} = 90$ ways

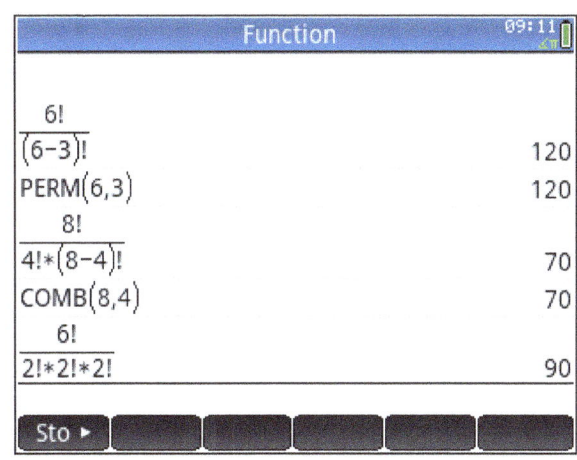

Find the number of ways to draw a 5-card hand form a deck of cards having two kings and three jacks?
$C(4,2) \cdot C(4,3) = 6 \cdot 4 = 24$

To meet the Political Science US government requirement a student needs eight out of ten true or false questions correct. How many ways can the student answer the exam to meet the requirement?
$C(10,8) + C(10,9) + C(10,10) = 45 + 10 + 1 = 56$

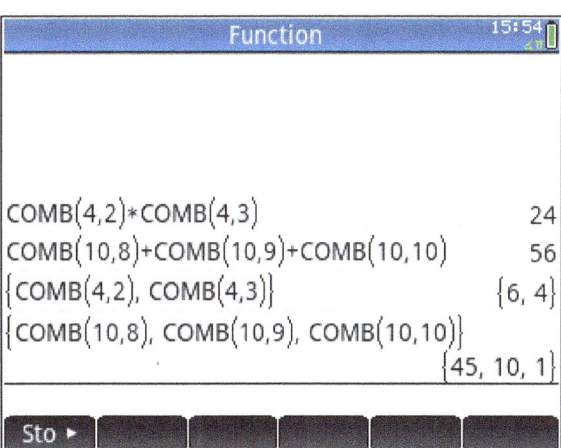

Experiment or Observation

Sunday, July 7, 2019   4:19 AM

http://computerlearningservice.com/Academy/Finite-Math/Sets-Probability/Experiment-or-Observation/experiment-or-observation.html

Terminology

Sunday, July 7, 2019   4:32 AM

Outcomes
Most basic observation or experiment result.

Sample Points
Individual outcome of sample space.

Sample Space
All possible outcomes.

Event
Any subset of the sample space. One individual outcome is a simple event.

Sample Space three coins tossed consist of eight possibilities.
S = {HHH, HHT, HTH, HTT, THH, THT, TTH, TTT}

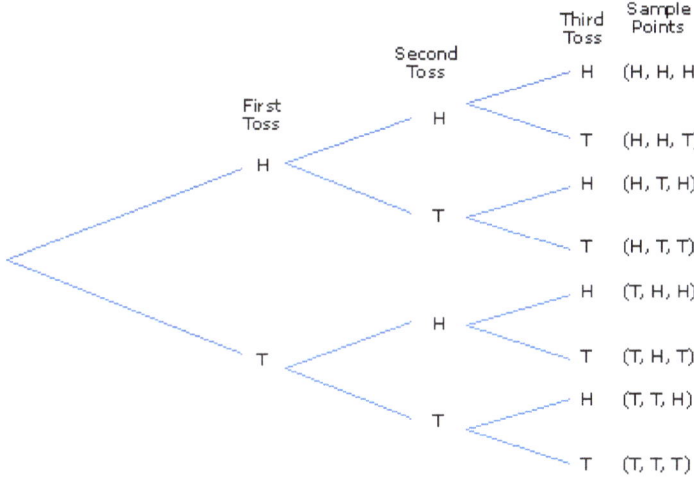

Event, exactly two tails
E_1 = {HTT, THT, TTH}

Event, at least two tails
E_2 = {HTT, THT, TTH, TTT}

An impossible event
∅

A certain event
S

Subpage

Sunday, July 7, 2019   12:34 PM

Success 7 or 11

S = {
    (1,1),(1,2),(1,3),(1,4),(1,5),(1,6)
    (2,1),(2,2),(2,3),(2,4),(2,5),(2,6)
    (3,1),(3,2),(3,3),(3,4),(3,5),(3,6)
    (4,1),(4,2),(4,3),(4,4),(4,5),(4,6)
    (5,1),(5,2),(5,3),(5,4),(5,5),(5,6)
    (6,1),(6,2),(6,3),(6,4),(6,5),(6,6)
    }

$E_7 \cup E_{11} =$
$\{(6,1), (5,2), (4,3), (3,4), (2,5), (1,6), (6,5), (5,6)\}$

At least 10

S = {
    (1,1),(1,2),(1,3),(1,4),(1,5),(1,6)
    (2,1),(2,2),(2,3),(2,4),(2,5),(2,6)
    (3,1),(3,2),(3,3),(3,4),(3,5),(3,6)
    (4,1),(4,2),(4,3),(4,4),(4,5),(4,6)
    (5,1),(5,2),(5,3),(5,4),(5,5),(5,6)
    (6,1),(6,2),(6,3),(6,4),(6,5),(6,6)
    }

$E_{10} \cup E_{11} \cup E_{12} =$
$\{(6,4), (5,5), (4,6), (6,5), (5,6), (6,6)\}$

Venn Diagrams

Sunday, July 7, 2019   8:20 AM

Union L1, L2

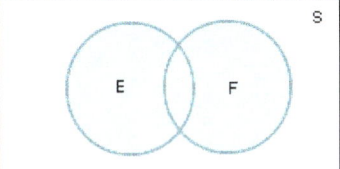

Diagram	HP Prime	Set Notation
E	L1	{2, 3, 4}
F	L2	{3, 4, 5}
$E \cup F$	$L1 \cup L2$	{2, 3, 4, 5}
$E \cap F$	$L1 \cap L2$	{3, 4}
$(E \cup F)^c$	L3	{1, 6}

$S = \{1, 2, 3, 4, 5, 6\}$

Intersection L1, L2

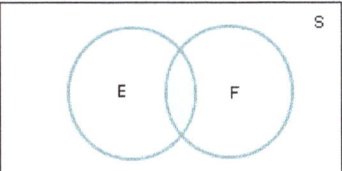

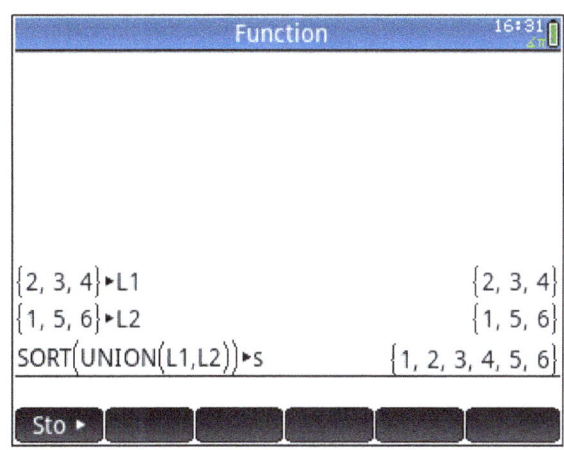

Complement E

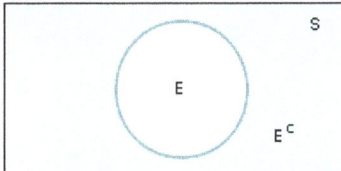

Diagram	HP Prime	Set Notation
E	L1	{2, 3, 4}
E^c	L2	{1, 5, 6}

$S = \{1, 2, 3, 4, 5, 6\}$

Mutually Exclusive L1, L2

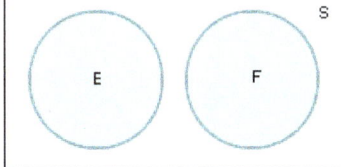

Diagram	HP Prime	Set Notation
E	L1	{1, 2}
F	L2	{3, 4}
$E \cup F$	$L1 \cup L2$	{1, 2, 3, 4}
$E \cap F$	$L1 \cap L2$	∅
$(E \cup F)^c$	L3	{5, 6}

$S = \{1, 2, 3, 4, 5, 6\}$

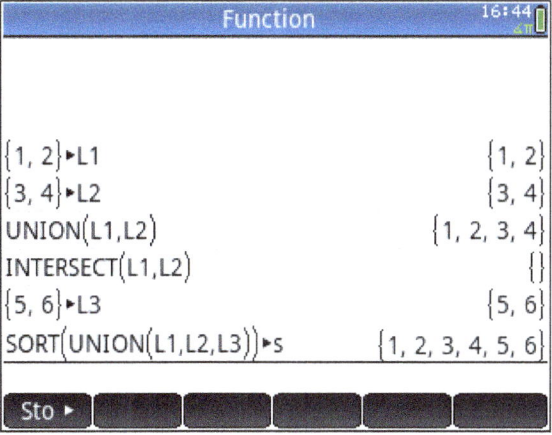

Sets and Probabilities Page 97

Probability

Monday, July 8, 2019   2:22 PM

http://computerlearningservice.com/Academy/Finite-Math/Sets-Probability/Probability/probability.html

Theoretical and Subjective

Monday, July 8, 2019   2:24 PM

Theoretical - Equally Likely Outcomes
Step 1. Count the total possible outcomes, $n(S)$.
Step 2. Count the total possible ways the item of interest occurs, $n(E)$.
Step 3. $P(E) = \dfrac{n(E)}{n(S)}$

Find the probability, exactly one tail when coin is tossed twice?
S = {HH, HT, TH, TT}, E = {HT, TH}

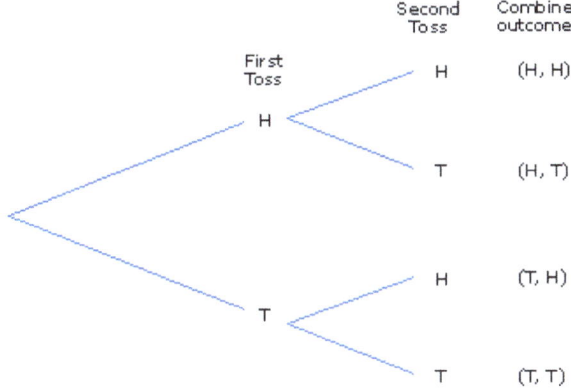

$P(E) = \dfrac{n(E)}{n(S)} = \dfrac{2}{4} = \dfrac{1}{2}$

Find the probability, getting a 2 when you roll a die?
S = {1, 2, 3, 4, 5, 6}, E = {2}

$P(E) = \dfrac{n(E)}{n(S)} = \dfrac{1}{6}$

Subjective
Estimate based on experience or intuition.
75% chance a friend will get engaged next year.

Additional Topics

Monday, July 8, 2019   3:21 PM

A Probability Distribution

Simple Event	Probability
1	$\frac{1}{6}$
2	$\frac{1}{6}$
3	$\frac{1}{6}$
4	$\frac{1}{6}$
5	$\frac{1}{6}$
6	$\frac{1}{6}$

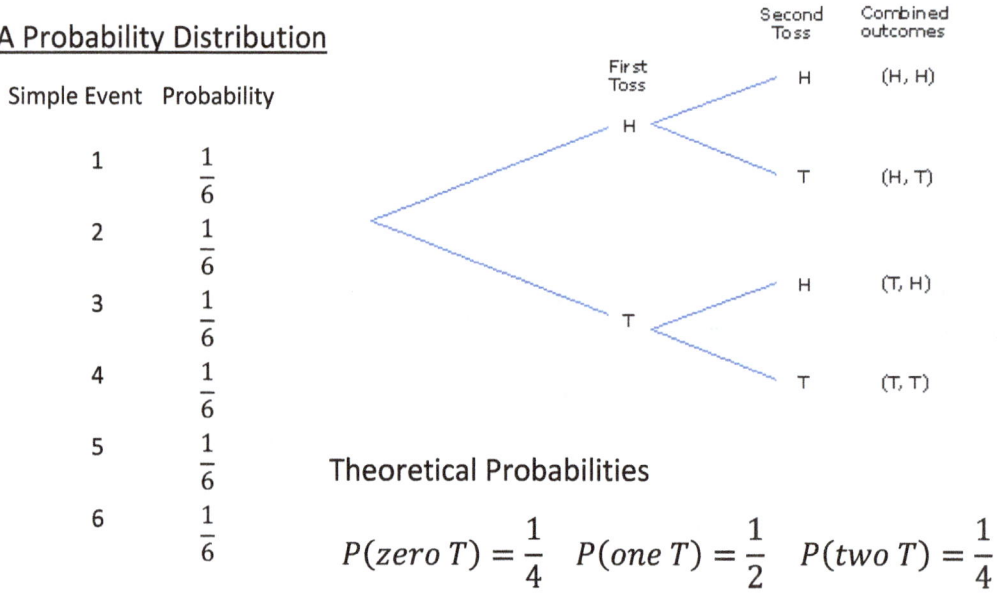

Theoretical Probabilities

$$P(zero\ T) = \frac{1}{4} \quad P(one\ T) = \frac{1}{2} \quad P(two\ T) = \frac{1}{4}$$

Relative Frequency

Step 1. Repeat or observe and count times event occurs, $n(E)$.
Step 2. Estimate $P(E)$ using

$$P(E) = \frac{n(E)}{n(observations)}$$

Repeat the coin tossed twice 100 times. Compare with above theoretical. Is the coin fair?

Zero tails 27 times. $P(E) = \frac{27}{100} = .27 \approx .25$

One tail 48 times. $P(E) = \frac{48}{100} = .48 \approx .50$

Two tails 25 times. $P(E) = \frac{25}{100} = .25$

Relative frequency is approximately same as theoretical. Coin is fair.

Rules of Probability

Tuesday, July 9, 2019   10:37 PM

http://computerlearningservice.com/Academy/Finite-Math/Sets-Probability/Rules-of-Probability/rules-of-probability.html

Basic Properties

Mutually Exclusive $E \cap F = \emptyset$

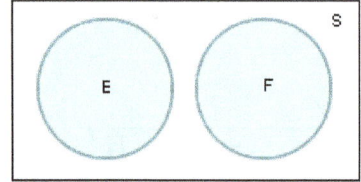

$E \cap B \neq \emptyset$

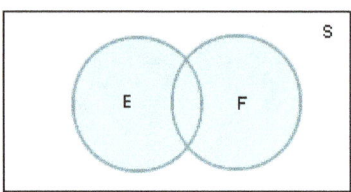

E^c Complement

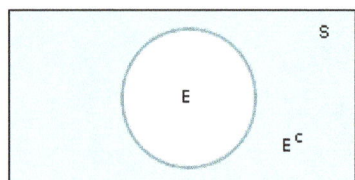

Basic Properties

$0 \leq P(E) \leq 1$ For all Event E

$P(S) = 1$ Sample Space S

$P(E \cup F) = P(E) + P(F)$ $E \cap F = \emptyset$

$P(E \cup F) = P(E) + P(F) - P(E \cap F)$ $E \cap B \neq \emptyset$

$P(E^c) = 1 - P(E)$ E^c Complement

A regular die is rolled. What is probability of rolling a 7?

$$P(E) = \frac{n(E)}{n(S)} = \frac{0}{6} = 0$$

A regular die is rolled. What is probability of rolling a 3 or 4?

$$P(E \cup F) = P(E) + P(F) = \frac{1}{6} + \frac{1}{6} = \frac{2}{6} = \frac{1}{3}$$

From a deck of cards, what is probability of drawing an ace or a spade?

$$P(E \cup F) = P(E) + P(F) - P(E \cap F) = \frac{4}{52} + \frac{13}{52} - \frac{1}{52}$$
$$= \frac{16}{52} = \frac{4}{13}$$

Computations

Tuesday, July 9, 2019   10:41 PM

Event E and F are mutually exclusive. $P(E) = .3$ and $P(F) = .5$.

$P(E \cap F) = 0$

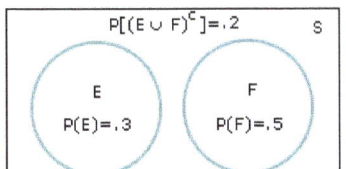

$P(E \cup F) = .3 + .5 = .8$

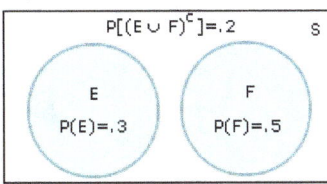

$P(E^c) = 1 - .3 = .7$

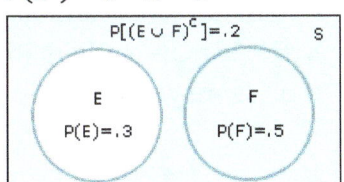

$P(E^c \cap F^c) = P(E \cup F)^c = .2$

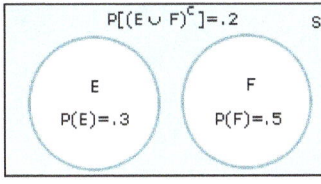

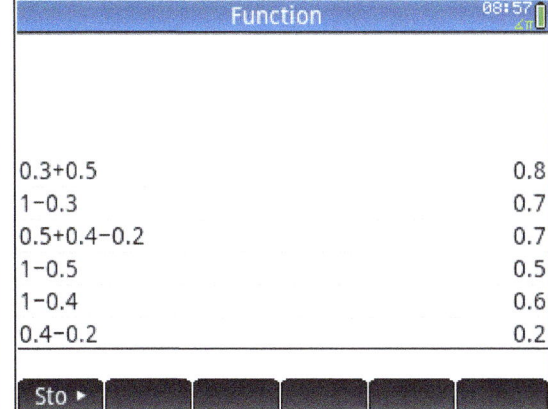

Event E and F have the following probabilities. $P(E) = .5$ and $P(F) = .4$ and $P(E \cap F) = .2$.

$P(E \cup F) = .5 + .4 - 2 = .7$

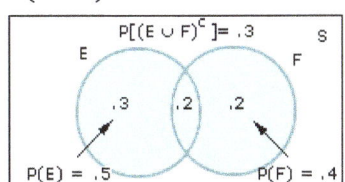

$P(E^c) = 1 - .5 = .5$

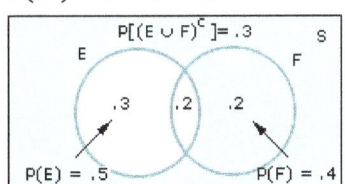

$P(F^c) = 1 - .4 = .6$

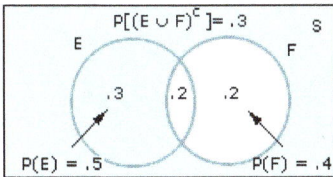

$P(E^c \cap F) = .4 - .2 = .2$

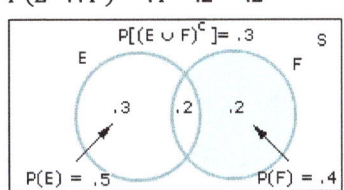

Sets and Probabilities Page 103

6

Additional Topics Probabilities

> Do not worry about your difficulties in mathematics. I can assure you mine are still greater."
> ~ Albert Einstein

Counting Techniques

- Additional Investigations   105
- Higher Sample Points   106

Conditional Probability

- Formulas   108
- Independent and Dependent Events   109

Bayes' Theorem

- Formula   112

Probability Distributions

- Random Variables   114
- Histogram   115
- 7 or 11   116

Expectations

- Expected Value   118
- Odds   119

Variance and Standard Deviation

- Manual   121
- App   122
- Chebychev's Inequality   123

Binomial Distributions

- Bernoulli Trails   125
- Formula   126

Normal Distribution Basics

- Density Curve   129
- Probabilities   130
- Z Scores   131

More Normal Distributions

- Other Means | Standard Deviations   133
- Approximate Binomial Distributions   135

Counting Techniques

Thursday, July 11, 2019   8:51 AM

http://computerlearningservice.com/Academy/Finite-Math/Additional-Topics/Counting-Techniques/counting-techniques.html

Additional Investigations

Thursday, July 11, 2019   8:56 AM

A coin is tossed five times. What is the probability lands tails exactly two times, at most two times, and on the first and fifth toss?

$$P(E_1) = \frac{n(E_1)}{n(S)} = \frac{C(5,2)}{2^5} = \frac{10}{32} = \frac{5}{16}$$

$$P(E_2) = \frac{n(E_2)}{n(S)} = \frac{C(5,0) + C(5,1) + C(5,2)}{2^5} = \frac{1+5+10}{32} = \frac{1}{2}$$

$$P(E_3) = \frac{n(E_3)}{n(S)} = \frac{1 \cdot 2 \cdot 2 \cdot 2 \cdot 1}{2^5} = \frac{8}{32} = \frac{1}{4}$$

A coin is tossed three times. What is the probability lands tails exactly two times, at most two times, and on the first and third toss?

S = {HHH, HHT, HTH, HTT, THH, THT, TTH, TTT}
E_1 = {HTT, THT, TTH}
E_2 = {HHH, HHT, HTH, THH, HTT, THT, TTH}
E_3 = {THT, TTT}

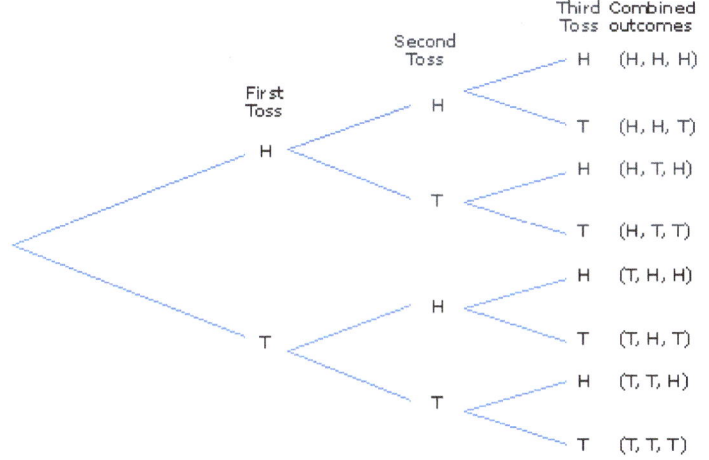

$$P(E_1) == \frac{n(E_1)}{n(S)} = \frac{C(3,2)}{2^3} = \frac{3}{8} \qquad P(E_3) = \frac{n(E_3)}{n(S)} = \frac{1 \cdot 2 \cdot 1}{2^3} = \frac{1}{4}$$

$$P(E_2) = \frac{n(E_2)}{n(S)} = \frac{C(3,0) + C(3,1) + C(3,2)}{2^3} = \frac{1+3+3}{8} = \frac{7}{8}$$

Higher Sample Points

Thursday, July 11, 2019   8:58 AM

A 2-person panel must be selected from a pool made up of 5 men and 7 women. What is the possibility of selecting an all male panel?

From a deck of 52 cards, the possibility that two red cards are drawn?

A jar has 25 yellow marbles, 32 green marbles, and 31 red marbles. Two marbles are drawn, what is probability one red and one yellow are drawn?

A jar has 25 yellow marbles, 32 green marbles, and 31 red marbles. Three marbles are drawn, what is probability of at least two green?

A pair of dice is rolled five times. For all five rolls what is probability of either doubles, seven, or eleven?

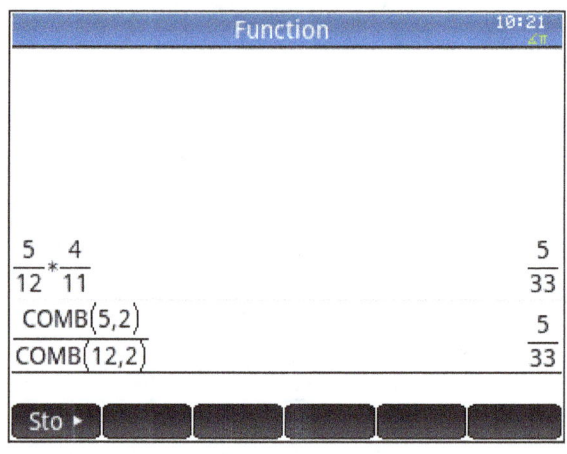

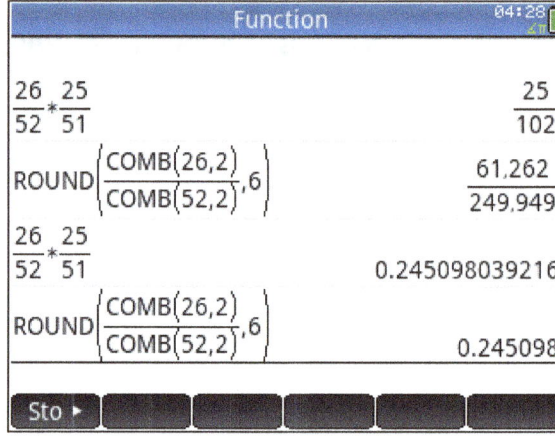

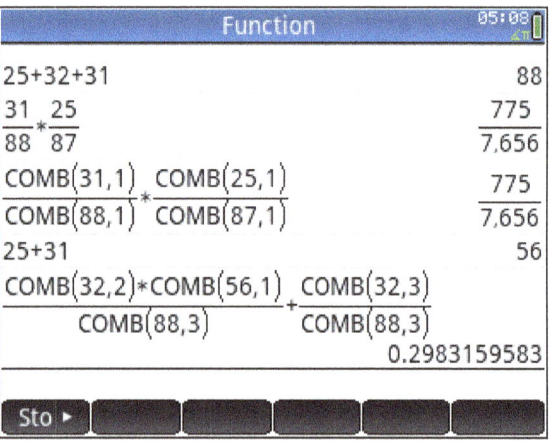

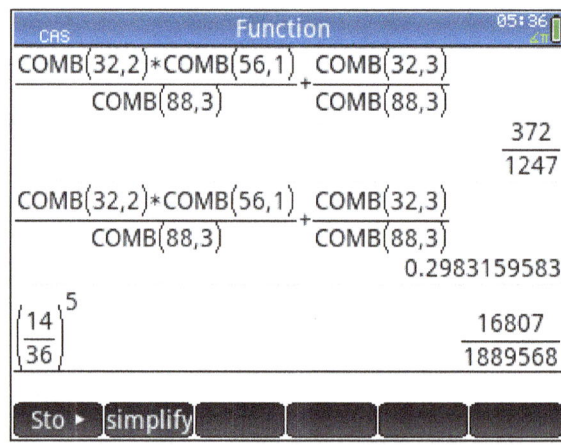

Additional Topics Probabilities Page 106

Conditional Probability

Wednesday, July 17, 2019   2:29 AM

http://computerlearningservice.com/Academy/Finite-Math/Additional-Topics/Conditional-Probability/conditional-probability.html

Formulas

Wednesday, July 17, 2019   2:30 AM

Conditional Formulas

$$P(E|F) = \frac{n(E \cap F)}{n(F)} = \frac{P(E \cap F)}{P(F)} \qquad \frac{n(E \cap F)}{n(F)} \cdot \frac{\frac{1}{n(s)}}{\frac{1}{n(s)}} = \frac{P(E \cap F)}{P(F)}$$

The $n(F) \neq 0$ and $P(F) \neq 0$.

A die is thrown. What is the probability four given even, three given even, and greater than two given odd?

$$P(E_1|F_1) = \frac{n(E_1 \cap F_1)}{n(F_1)} = \frac{n(\{4\}) \cap n(\{2,4,6\})}{n(\{2,4,6\})} = \frac{1}{3}$$

$$P(E_1|F_1) = \frac{P(E_1 \cap F_1)}{P(F_1)} = \frac{P(\{4\}) \cap P(\{2,4,6\})}{P(\{2,4,6\})} = \frac{\frac{1}{6}}{\frac{3}{6}} = \frac{1}{3}$$

$$P(E_2|F_2) = \frac{n(E_2 \cap F_2)}{n(F_2)} = \frac{n(\{3\}) \cap n(\{2,4,6\})}{n(\{2,4,6\})} = \frac{0}{3} = 0$$

$$P(E_2|F_2) = \frac{P(E_2 \cap F_2)}{P(F_2)} = \frac{P(\{3\}) \cap P(\{2,4,6\})}{P(\{2,4,6\})} = \frac{\frac{0}{6}}{\frac{3}{6}} = \frac{0}{3} = 0$$

$$P(E_3|F_3) = \frac{n(E_3 \cap F_3)}{n(F_3)} = \frac{n(\{3,4,5,6\}) \cap n(\{1,3,5\})}{n(\{1,3,5\})} = \frac{2}{3}$$

$$P(E_3|F_3) = \frac{P(E_3 \cap F_3)}{P(F_3)} = \frac{P(\{3,4,5,6\}) \cap P(\{1,3,5\})}{P(\{1,3,5\})} = \frac{\frac{2}{6}}{\frac{3}{6}} = \frac{2}{3}$$

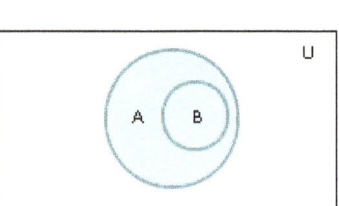

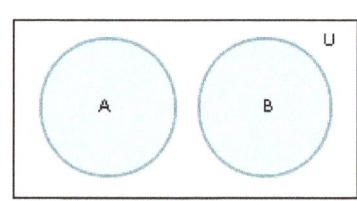

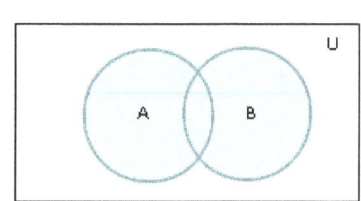

Independent and Dependent Events

Independent Events
The occurrence of one event does not affect outcome of the probability of the second event. The probability both occur is
$P(E \cap F) = P(E) \cdot P(F)$ since $P(E|F) = P(E)$

Dependent Events
The occurrence of one event changes the probability of the second event. The probability both occur is
$P(E \cap F) = P(E) \cdot P(F|E)$ using $P(F|E)$ in our Conditional Formula

A coin is tossed twice. Show that Heads in a second toss is independent of Tails in the first toss, the {T, H} branch of the tree.

Using abbreviated form we have:

S={HH, HT, TH, TT}, E={TH, TT}, F={HH, TH}, $E \cap F$={TH}
$P(E) = \frac{1}{2}, P(F) = \frac{1}{2}, P(E \cap F) = \frac{1}{4}$
$P(E \cap F) = P(E) \cdot P(F) = \frac{1}{2} \cdot \frac{1}{2} = \frac{1}{4}$

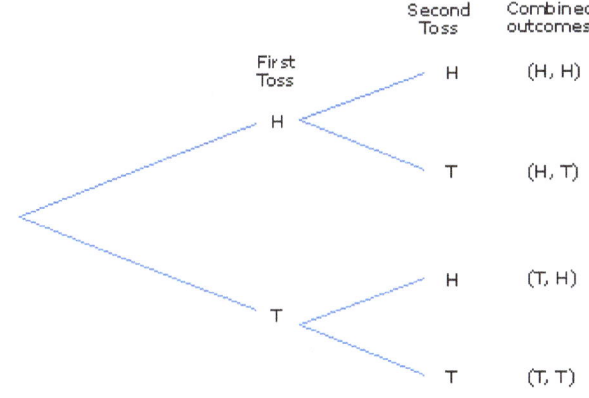

Subpage

Thursday, July 18, 2019   3:36 AM

Are the events drawing a King, *E*, and drawing a Black card, *F*, independent? Answer -Yes

$P(E) = \frac{4}{52} = \frac{1}{13}, P(F) = \frac{26}{52} = \frac{1}{2}, P(E \cap F) = \frac{2}{52} = \frac{1}{26}$

$P(E \cap F) = P(E) \cdot P(F) = \frac{1}{13} \cdot \frac{1}{2} = \frac{1}{26}$

Finite Stochastic Process

Two cards are drawn without replacement. What is the probability that a face card is drawn on a second draw? (Draw the tree. Assume first card is a face card.)

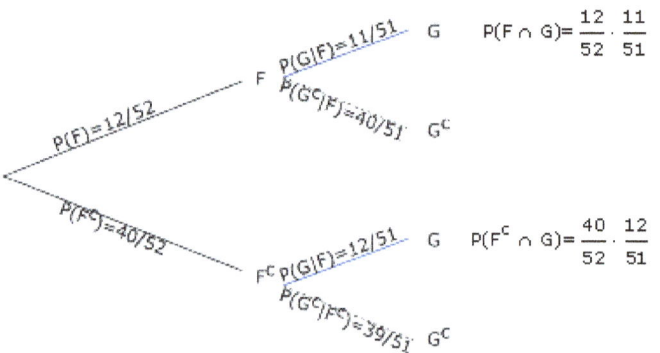

$T = M_2 = \begin{bmatrix} G|F & G|F^c \\ G^c|F & G^c|F^c \end{bmatrix} \quad X_0 = M_1 = \begin{bmatrix} F \\ F^c \end{bmatrix}$

$M_2 \cdot M_1 = \begin{bmatrix} G \cap F + G \cap F^c \\ G^c \cap F + G^c \cap F^c \end{bmatrix} = \begin{bmatrix} G \\ G^c \end{bmatrix}$

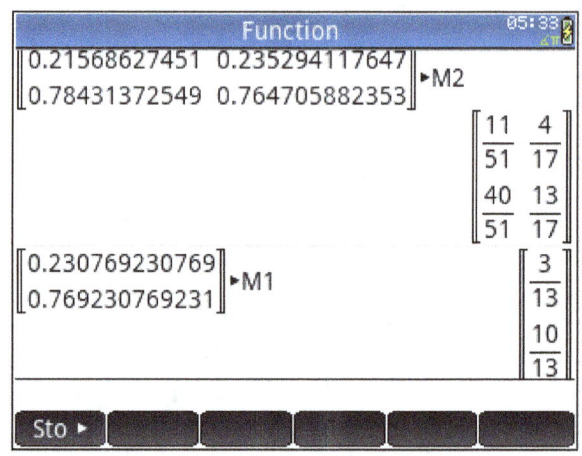

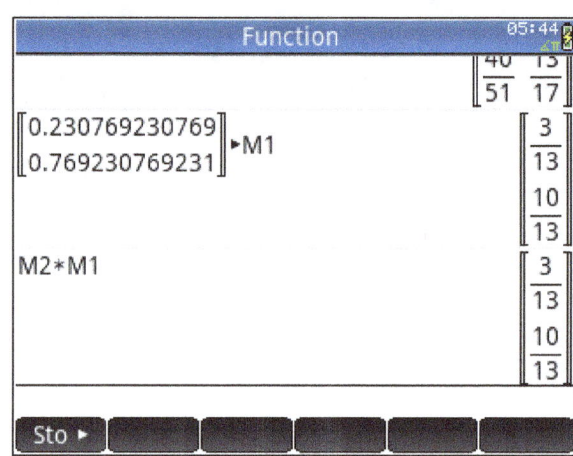

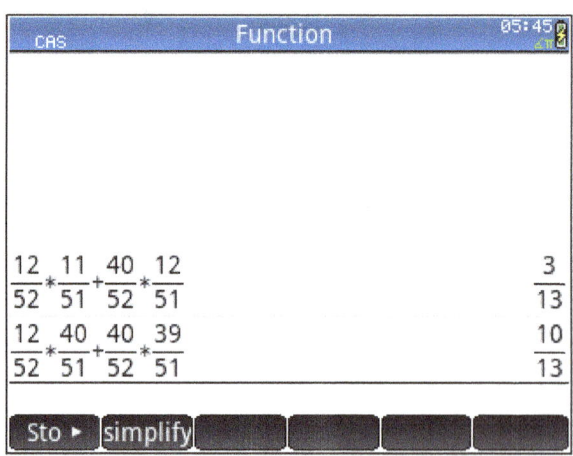

Bayes' Theorem

Friday, August 2, 2019   2:07 AM

http://computerlearningservice.com/Academy/Finite-Math/Additional-Topics/Bayes--Theorem/bayes--theorem.html

Formula

Friday, August 2, 2019   4:10 AM

$$P(B|D) = \frac{P(B) \cdot P(D|B)}{P(A) \cdot P(D|A) + P(B) \cdot P(D|B) + P(C) \cdot P(D|C)}$$

$$P(B|D) = \frac{(.43) \cdot (.48)}{(.26) \cdot (.68) + (.43) \cdot (.48) + (.31) \cdot (.20)} = \frac{.2064}{.4452} \approx .464$$

$$T = M_2 = \begin{bmatrix} .68 & .48 & .20 \\ .32 & .52 & .80 \end{bmatrix}$$

$$X_0 = M_1 = \begin{bmatrix} .26 \\ .43 \\ .31 \end{bmatrix}$$

$$M_3 = M_2 \cdot M_1 = T \cdot X_0 = \begin{bmatrix} .4452 \\ .5548 \end{bmatrix}$$

$$P(B|D) = \frac{M1(2,1) \cdot M2(1,2)}{M3(1,1)} = \frac{(.43) \cdot (.48)}{.4452} = \frac{.2064}{.4452} \approx .464$$

$$P(A_i|E) = \frac{n(A_i \cap E)}{n(E)} = \frac{P(A_i \cap E)}{P(E)}$$

$$= \frac{P(A_i \cap E)}{P(A_1 \cap E) + P(A_2 \cap E) + \cdots + P(A_n \cap E)}$$

Bayes' Formula
Let S be a sample space with A_i being n partitions of S.

$$P(A_i|E) = \frac{P(A_i) \cdot P(E|A_i)}{P(A_1) \cdot P(E|A_1) + P(A_2) \cdot P(E|A_2) + \cdots + P(A_2) \cdot P(E|A_n)}$$

Probability Distributions

Sunday, August 4, 2019   2:25 PM

http://computerlearningservice.com/Academy/Finite-Math/Additional-Topics/Probability-Distributions/probability-distributions.html

Random Variables

Sunday, August 4, 2019   2:26 PM

Random Variables
A random variable with an experiment with sample space S is a rule that assigns a number to each outcome.

A coin is tossed three times. Let the random variable X represent the number of tails that occur.

$S = \{HHH, HHT, HTH, HTT, THH, THT, TTH, TTT\}$

Event Distribution

Outcomes	HHH	HHT	HTH	HTT	THH	THT	TTH	TTT
X	0	1	1	2	1	2	2	3

The event ($X = 2$) is { HTT, THT, TTH}. $P(X = 2) = \dfrac{3}{8}$

Probability Distribution

X	0	1	2	3
P(X = x)	$\dfrac{1}{8}$	$\dfrac{3}{8}$	$\dfrac{3}{8}$	$\dfrac{1}{8}$

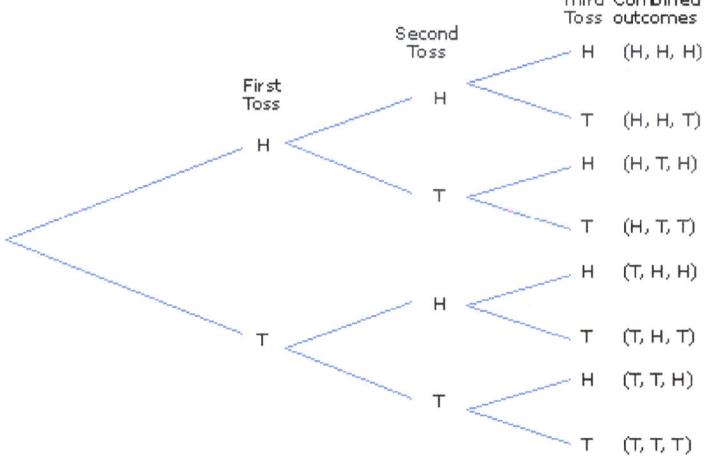

Histograms

Sunday, August 4, 2019   2:27 PM

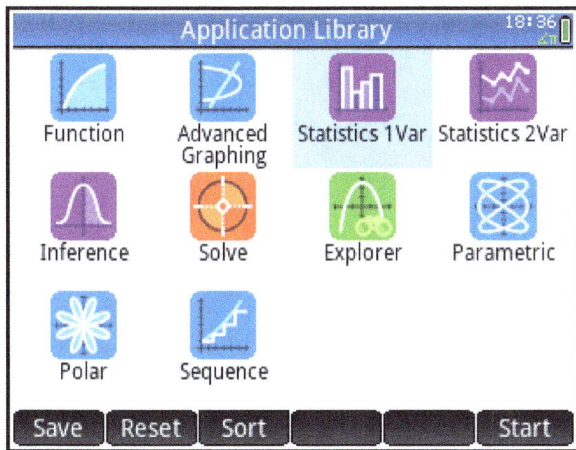

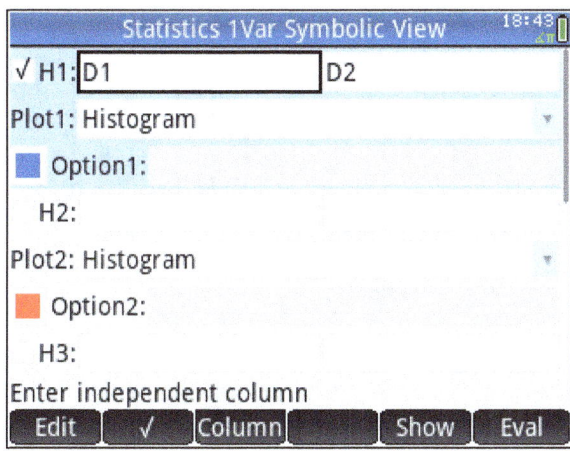

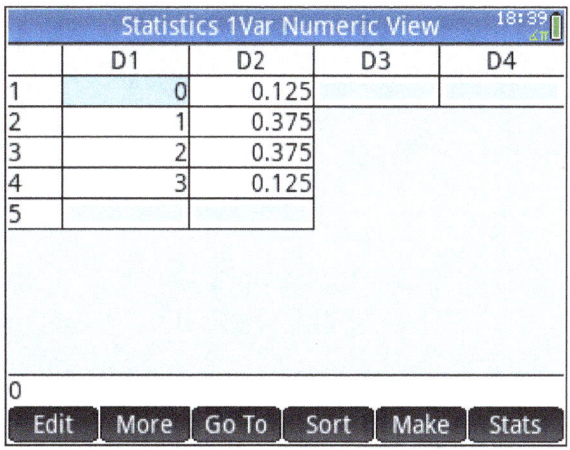

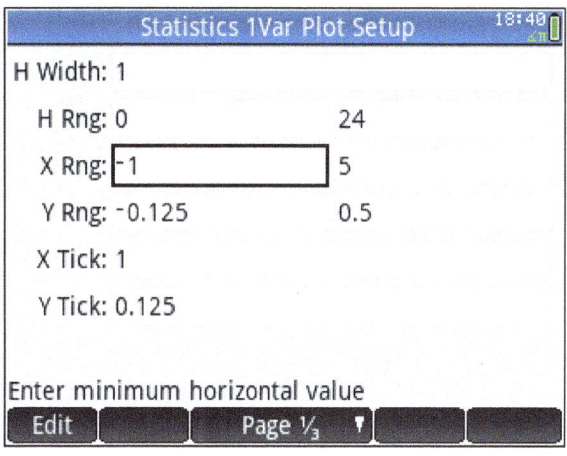

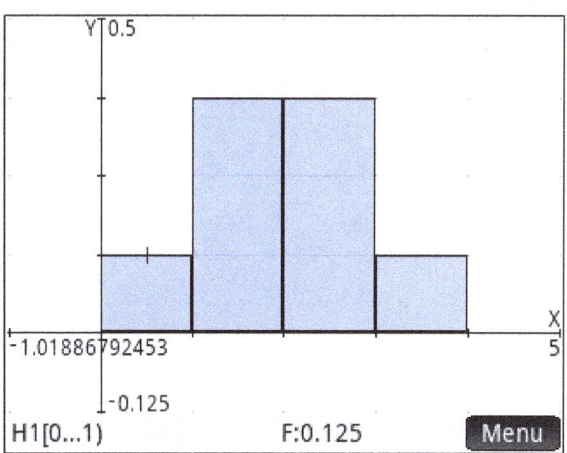

7 or 11

Monday, August 5, 2019   4:09 AM

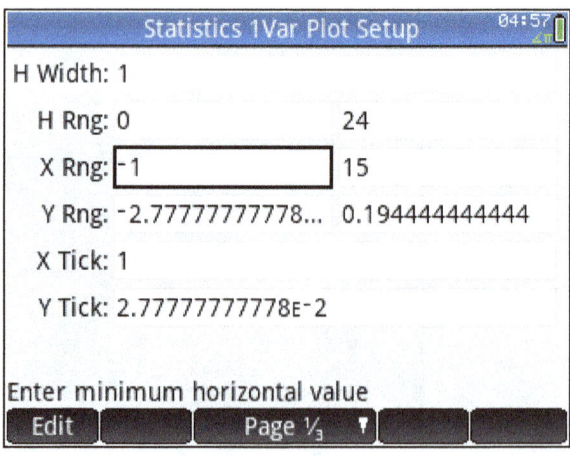

Success 7 or 11

S = {
(1,1),(1,2),(1,3),(1,4),(1,5),(1,6)
(2,1),(2,2),(2,3),(2,4),(2,5),(2,6)
(3,1),(3,2),(3,3),(3,4),(3,5),(3,6)
(4,1),(4,2),(4,3),(4,4),(4,5),(4,6)
(5,1),(5,2),(5,3),(5,4),(5,5),(5,6)
(6,1),(6,2),(6,3),(6,4),(6,5),(6,6)
}

$(X = 7 \cup X = 11) =$

$\{(6,1), (5,2), (4,3), (3,4), (2,5), (1,6), (6,5), (5,6)\}$

$P(X = 7 \cup X = 11) = \dfrac{6}{36} + \dfrac{2}{36} = \dfrac{8}{36} = \dfrac{2}{9}$

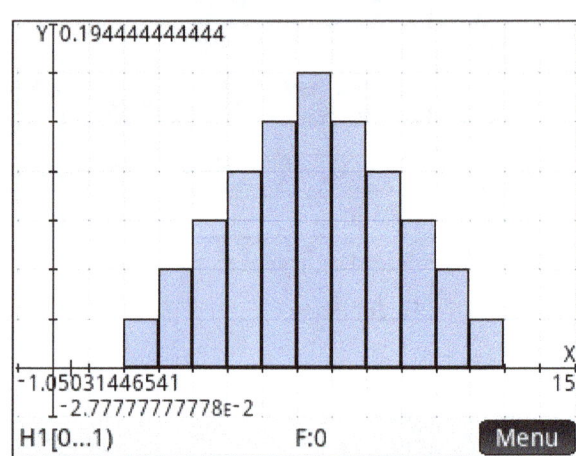

Probability Distribution

X	2	3	4	5	6	7	8	9	10	11	12
P(X = x)	$\dfrac{1}{36}$	$\dfrac{2}{36}$	$\dfrac{3}{36}$	$\dfrac{4}{36}$	$\dfrac{5}{36}$	$\dfrac{6}{36}$	$\dfrac{5}{36}$	$\dfrac{4}{36}$	$\dfrac{3}{36}$	$\dfrac{2}{36}$	$\dfrac{1}{36}$

Expectations

Sunday, August 11, 2019   2:08 AM

http://computerlearningservice.com/Academy/Finite-Math/Additional-Topics/Expectations/expectations.html

Expected Value

Sunday, August 11, 2019   2:10 AM

Mean or Average
The sum of all values divided by total number of values.

Finance Exam Distribution

Scores	95-99	90-94	85-89	80-84	75-79	70-74	65-69	60-64
X-Representative	97	92	87	82	77	72	67	62
Frequency	4	4	8	8	4	4	0	0
Probability	$\frac{4}{32}$	$\frac{4}{32}$	$\frac{8}{32}$	$\frac{8}{32}$	$\frac{4}{32}$	$\frac{4}{32}$	$\frac{0}{32}$	$\frac{0}{32}$

$$\frac{97 \cdot 4 + 92 \cdot 4 + 87 \cdot 8 + 82 \cdot 8 + 77 \cdot 4 + 72 \cdot 4}{32} = 84.5$$

Expected Value
Let X be a random variable, x_i be the number assigned by X.
$$E(X) = x_1 \cdot P(X = x_1) + x_2 \cdot P(X = x_2) + \cdots + x_n \cdot P(X = x_n)$$

$$97 \cdot \frac{4}{32} + 92 \cdot \frac{4}{32} + 87 \cdot \frac{8}{32} + 82 \cdot \frac{8}{32} + 77 \cdot \frac{4}{32} + 72 \cdot \frac{4}{32} = 84.5$$

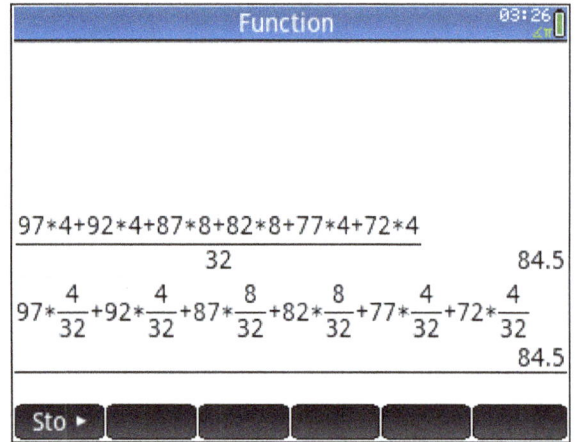
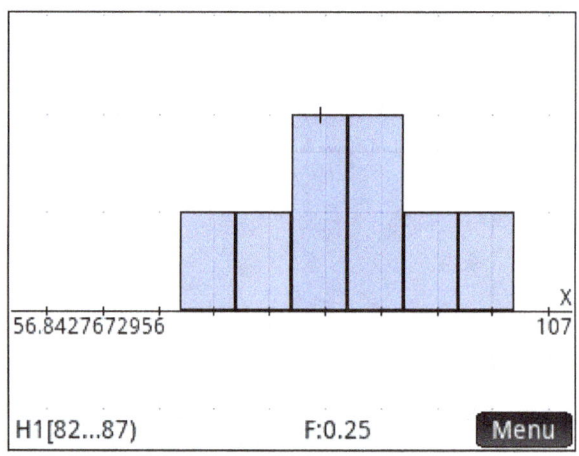

Odds

Monday, August 5, 2019   4:09 AM

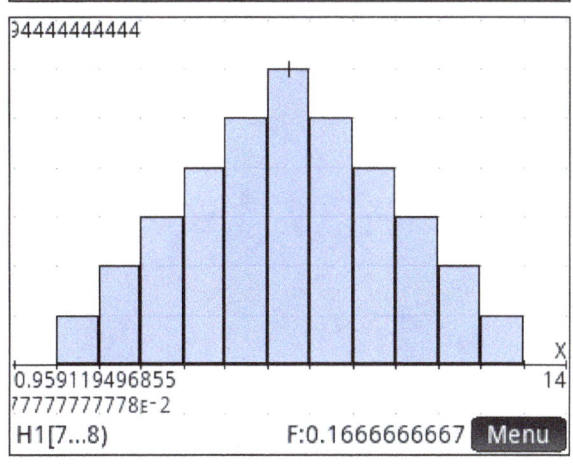

Success 7 or 11

S = {
   (1,1),(1,2),(1,3),(1,4),(1,5),(1,6)
   (2,1),(2,2),(2,3),(2,4),(2,5),(2,6)
   (3,1),(3,2),(3,3),(3,4),(3,5),(3,6)
   (4,1),(4,2),(4,3),(4,4),(4,5),(4,6)
   (5,1),(5,2),(5,3),(5,4),(5,5),(5,6)
   (6,1),(6,2),(6,3),(6,4),(6,5),(6,6)
}

$E = (X = 7 \cup X = 11) =$

$\{(6,1), (5,2), (4,3), (3,4), (2,5), (1,6), (6,5), (5,6)\}$

$$P(E) = P(X = 7 \cup X = 11) = \frac{6}{36} + \frac{2}{36} = \frac{8}{36} = \frac{2}{9}$$

Odds
For

$$\frac{P(E)}{P(E^c)} = \frac{P(E)}{1 - P(E)} = \frac{\frac{2}{9}}{\frac{7}{9}} = \frac{2}{7}$$

Against

$$\frac{P(E^c)}{P(E)} = \frac{1 - P(E)}{P(E)} = \frac{\frac{7}{9}}{\frac{2}{9}} = \frac{7}{2}$$

Probability Distribution

x	2	3	4	5	6	7	8	9	10	11	12
P(X = x)	$\frac{1}{36}$	$\frac{2}{36}$	$\frac{3}{36}$	$\frac{4}{36}$	$\frac{5}{36}$	$\frac{6}{36}$	$\frac{5}{36}$	$\frac{4}{36}$	$\frac{3}{36}$	$\frac{2}{36}$	$\frac{1}{36}$

Variance and Standard Deviation

Monday, August 12, 2019   10:09 AM

http://computerlearningservice.com/Academy/Finite-Math/Additional-Topics/Variance-Std-Dev/variance-std-dev.html

Manual

Monday, August 12, 2019   10:12 AM

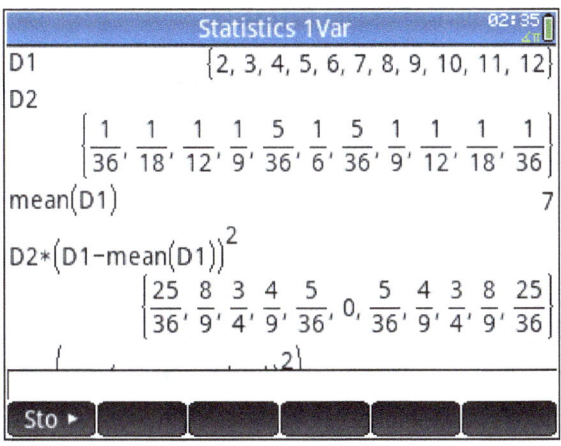
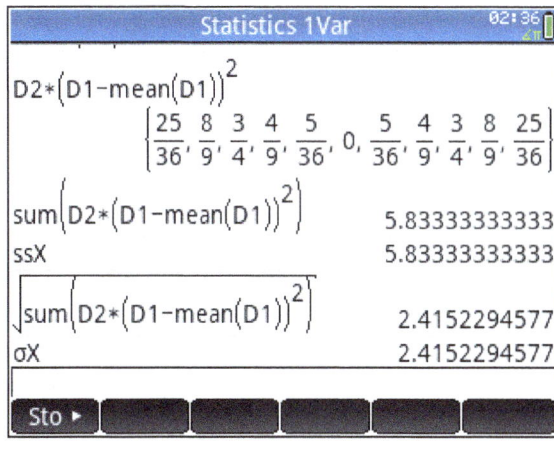
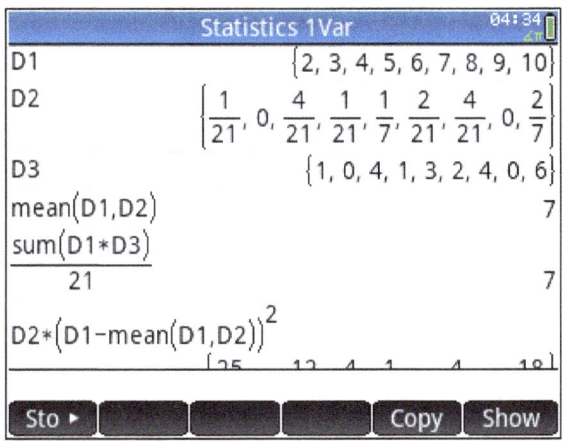
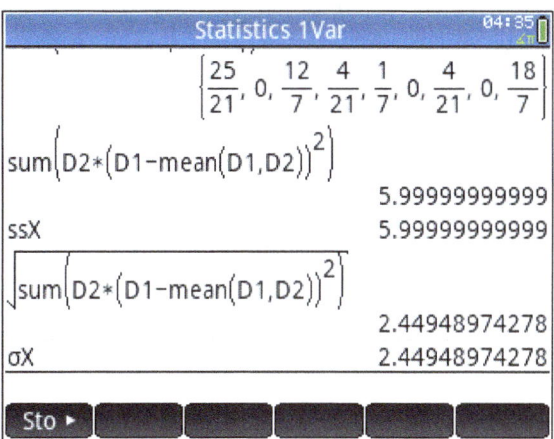

Variance

Let X be a random variable, x_i be the number assigned by X, with the probability p_i.

$$mean = \mu = E(X) \cdots \langle x\ bar \rangle$$

$$Var(X) = ssX = p_1 \cdot (x_1 - \mu)^2 + p_2 \cdot (x_2 - \mu)^2 + \cdots + p_n \cdot (x_n - \mu)^2$$

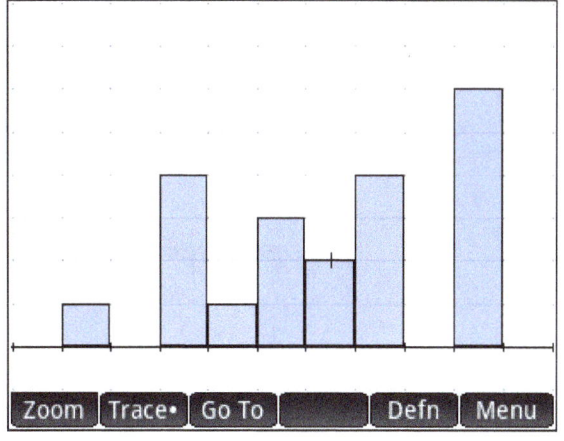

App

Monday, August 12, 2019   10:13 AM

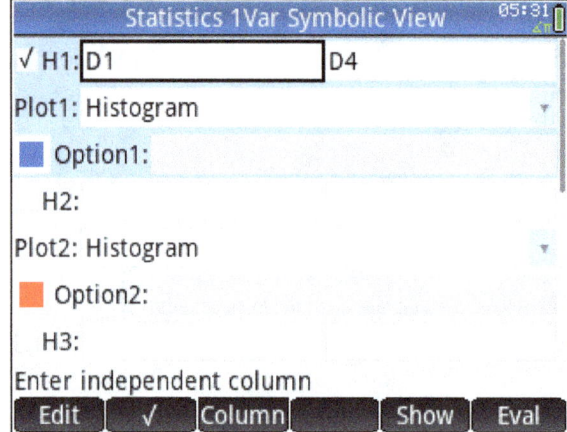
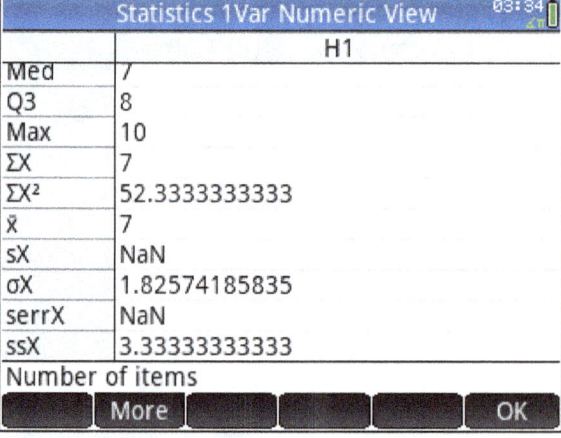
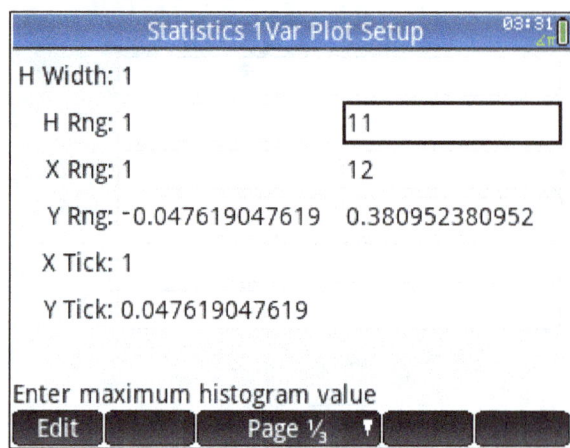
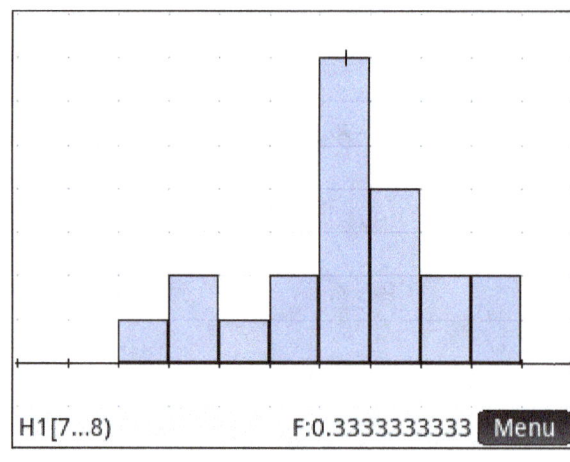

Standard Deviation

Let X be a random variable, x_i be the number assigned by X, with the probability p_i.

$$StdDev = \sigma = \sigma X = \sqrt{Var(X)}$$

Additional Topics Probabilities Page 122

Chebychev's Inequality

Thursday, August 15, 2019   3:12 AM

Chebychev's Inequality

Let X be a random variable, with expected value, $mean = \mu = E(X) \cdots \langle x\ bar \rangle$, and standard deviation, $StdDev = \sigma = \sigma X$, the probability is:

$$P(\mu - k\sigma \leq X \leq \mu + k\sigma) \geq 1 - \frac{1}{k^2}$$

A probability distribution has a mean of 7. Use Chebychev's Inequality to estimate the probability that outcome of quiz 2 is between 4 and 10. Answer 62.96% ≈ 63%

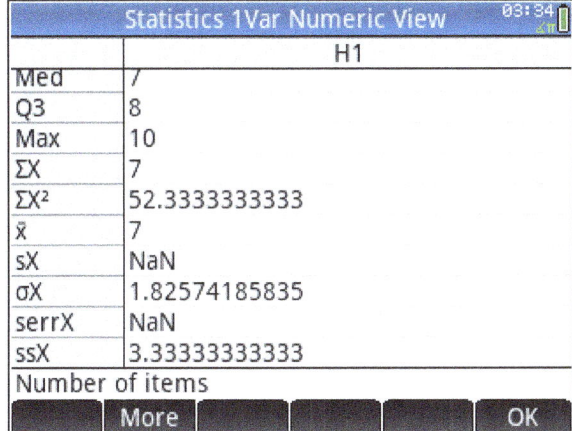

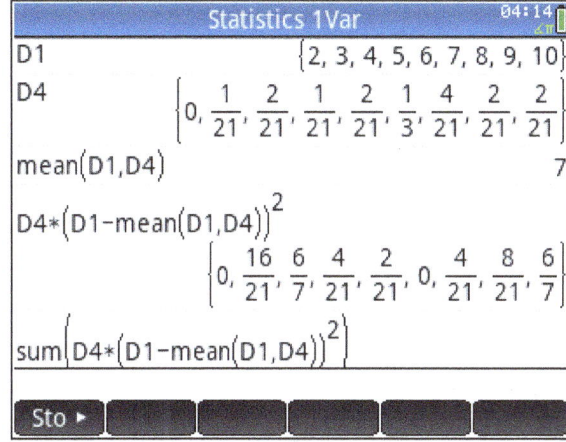

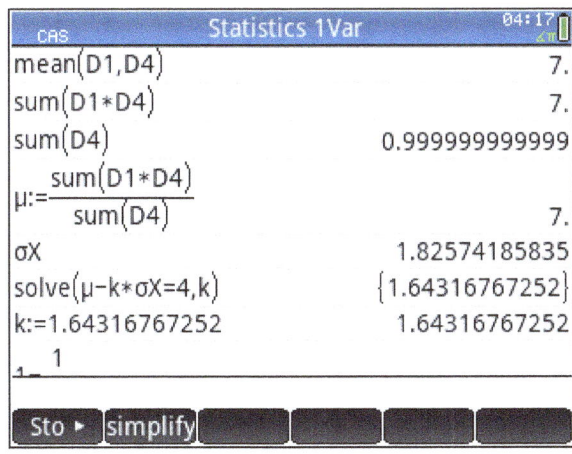

Additional Topics Probabilities Page 123

Binomial Distributions

Friday, August 16, 2019   2:16 AM

http://computerlearningservice.com/Academy/Finite-Math/Additional-Topics/Binomial-Distr/binomial-distr.html

Bernoulli Trials

Friday, August 16, 2019   2:17 AM

Bernoulli Trial
A Bernoulli Trial or binomial trial is any experiment that has two outcomes, success and failure. The trial is independent of previous trials of the experiment.

Four dice are thrown. What is probability that exactly one two comes up?

0 successes	1 successes	2 successes	3 successes	4 successes
FFFF	SFFF	SSFF	SSSF	SSSS
	FSFF	FSSF	FSSS	
	FFSF	FFSS	SSFS	
	FFFS	FSSF	SFSS	
		SFFS		
		FSFS		
$P(0) = q \cdot q \cdot q \cdot q$	$P(1) = p \cdot q \cdot q \cdot q$	$P(2) = p \cdot p \cdot q \cdot q$	$P(3) = p \cdot p \cdot p \cdot q$	$P(4) = p \cdot p \cdot p \cdot p$

Probability of 1 success, $p \cdot q \cdot q \cdot q$.
$$\frac{1}{6} \cdot \frac{5}{6} \cdot \frac{5}{6} \cdot \frac{5}{6} = \frac{125}{1296}$$

We have 4 Bernoulli trials
$$P(\text{exactly 2}) = 4 \cdot p \cdot q^3 = 4 \cdot \frac{125}{1296} \approx 0.386$$

Formula

Friday, August 16, 2019   2:41 AM

Formula
Bernoulli trial experiment the probability of r successes, p, given n trials.
$$P(r) = C(n,r)p^r q^{n-r} \quad \text{(where } q = 1-p\text{)}$$
The term $C(n,r) = \dfrac{n!}{r!(n-r)!}$ is known as the binomial coefficient.

BINOMIAL(n,p,r):=COMB(n,r)*p^r*(1-p)^(n-r)

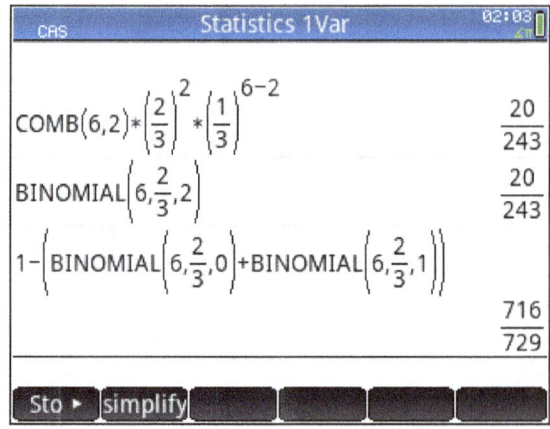

Six dice are thrown. What is probability that exactly two of the dice come up with at least three?

We have 6 trials. Probability of success, is a 3, 4, 5, or 6.
$$\frac{1}{6} + \frac{1}{6} + \frac{1}{6} + \frac{1}{6} = \frac{4}{6} = \frac{2}{3}$$

We have 6 trials choose 2, that is 15 in some order, of the product of
$$\frac{2}{3} \cdot \frac{2}{3} \cdot \frac{1}{3} \cdot \frac{1}{3} \cdot \frac{1}{3} \cdot \frac{1}{3} = \frac{4}{729}$$

$$P(\text{exactly 2}) = C(6,2) \cdot \left(\frac{2}{3}\right)^2 \left(\frac{1}{3}\right)^{6-2} = 15 \cdot \frac{4}{729} = \frac{20}{243} \approx 0.0823$$

Six dice are thrown. What is probability that at least two of the dice come up with at least three?

$$P(\text{at least 2}) = 1 - P(\text{at most 1}) =$$
$$1 - \left(C(6,0) \cdot \left(\frac{2}{3}\right)^0 \left(\frac{1}{3}\right)^{6-0} + C(6,1) \cdot \left(\frac{2}{3}\right)^1 \left(\frac{1}{3}\right)^{6-1} \right) = \frac{716}{729} \approx 0.9822$$

Subpage

Friday, August 16, 2019   6:29 AM

Formulas Binomial Distribution

$mean = \mu = E(X) = np \cdots \langle x\ bar \rangle$

$Var(X) = ssX = npq$

$StdDev = \sigma = \sigma X = \sqrt{npq}$

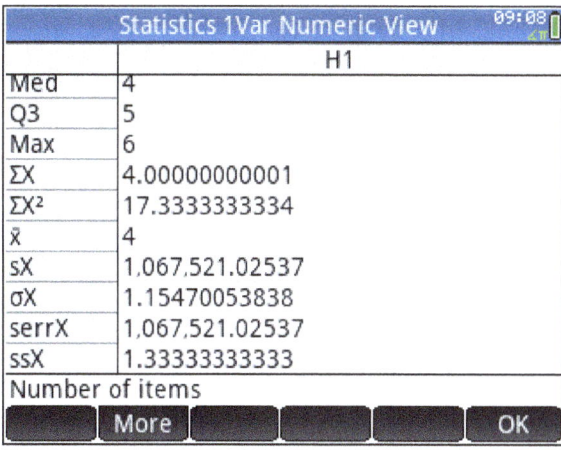

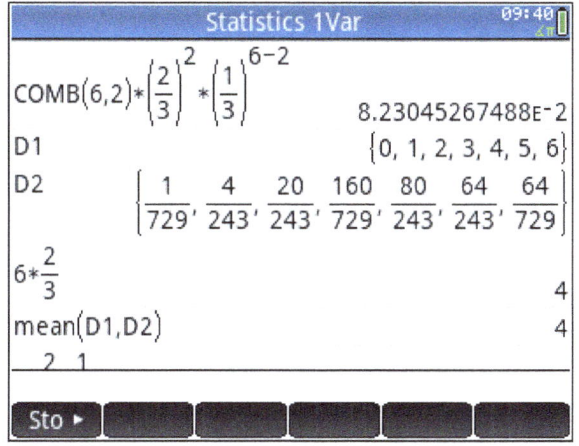

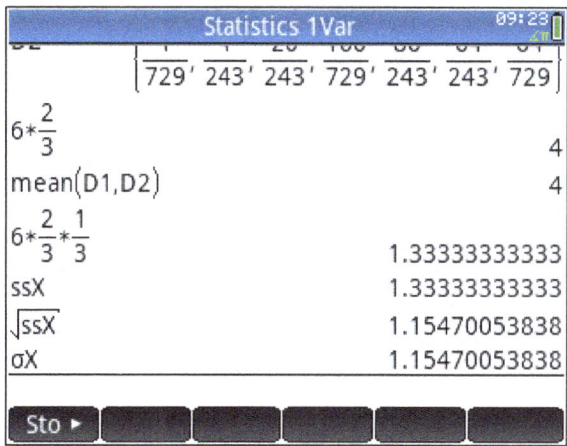

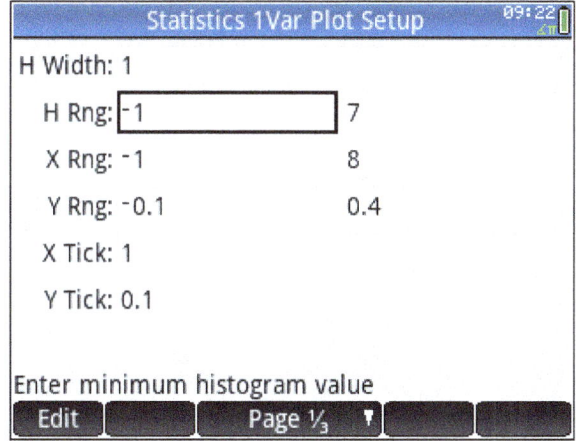

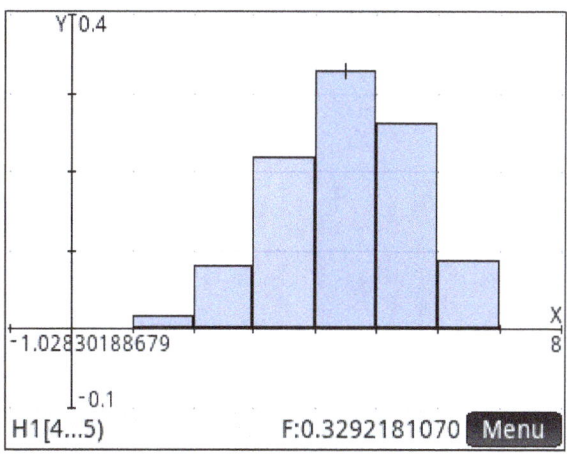

Additional Topics Probabilities Page 127

Normal Distribution Basics

Sunday, August 18, 2019   10:10 AM

http://computerlearningservice.com/Academy/Finite-Math/Additional-Topics/Normal-Distribution/normal-distribution.html

Density Curve

Sunday, August 18, 2019   10:12 AM

Probability Density Function Normal Curve
The Probability Density Function for the normal curve,

$$f(x) = \frac{1}{\sigma\sqrt{2\pi}} e^{-(1/2)[(x-\mu)/\sigma]^2} \qquad Z = \frac{X-\mu}{\sigma}$$

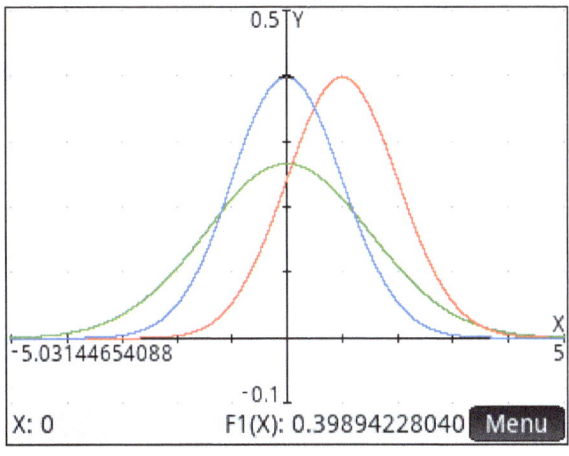

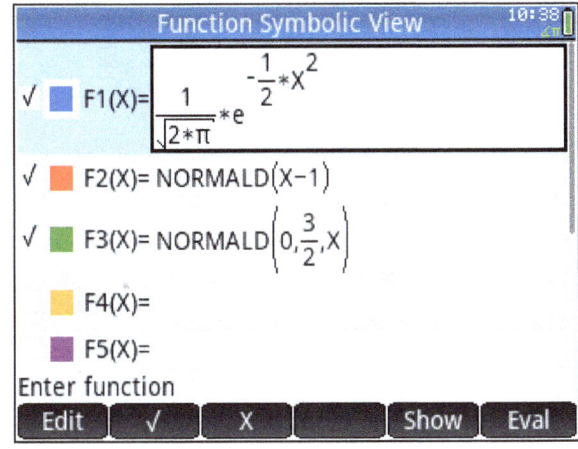

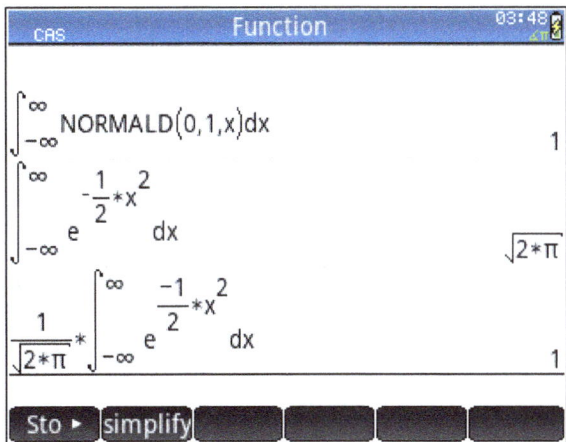

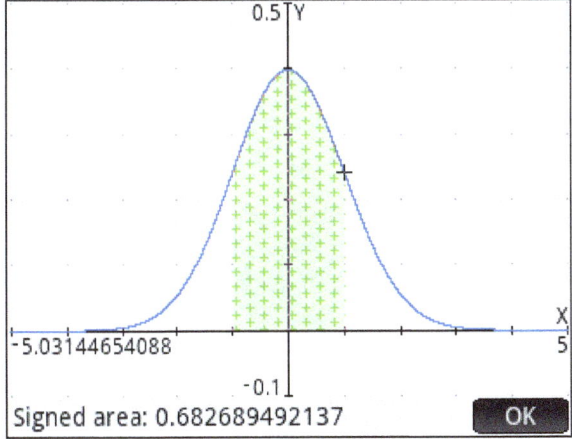

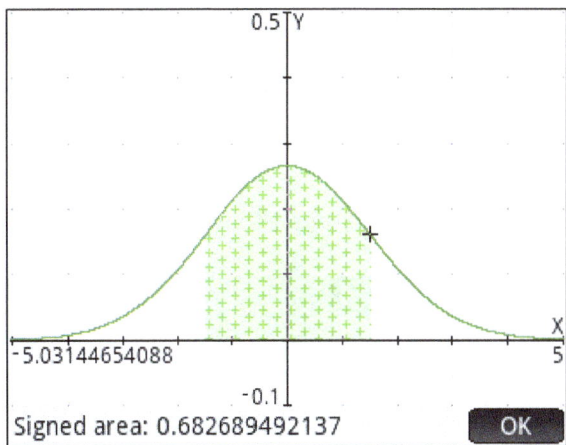

Additional Topics Probabilities Page 129

Probabilities

Sunday, August 18, 2019   10:12 AM

Plot the area under the normal curve corresponding to the probability. Find the probability Z value for the following:

$P(Z < 1.15) \quad P(Z > 0.81) \quad P(0.32 < Z < 2.14)) \quad P(-1.35 < Z < 1.85)$

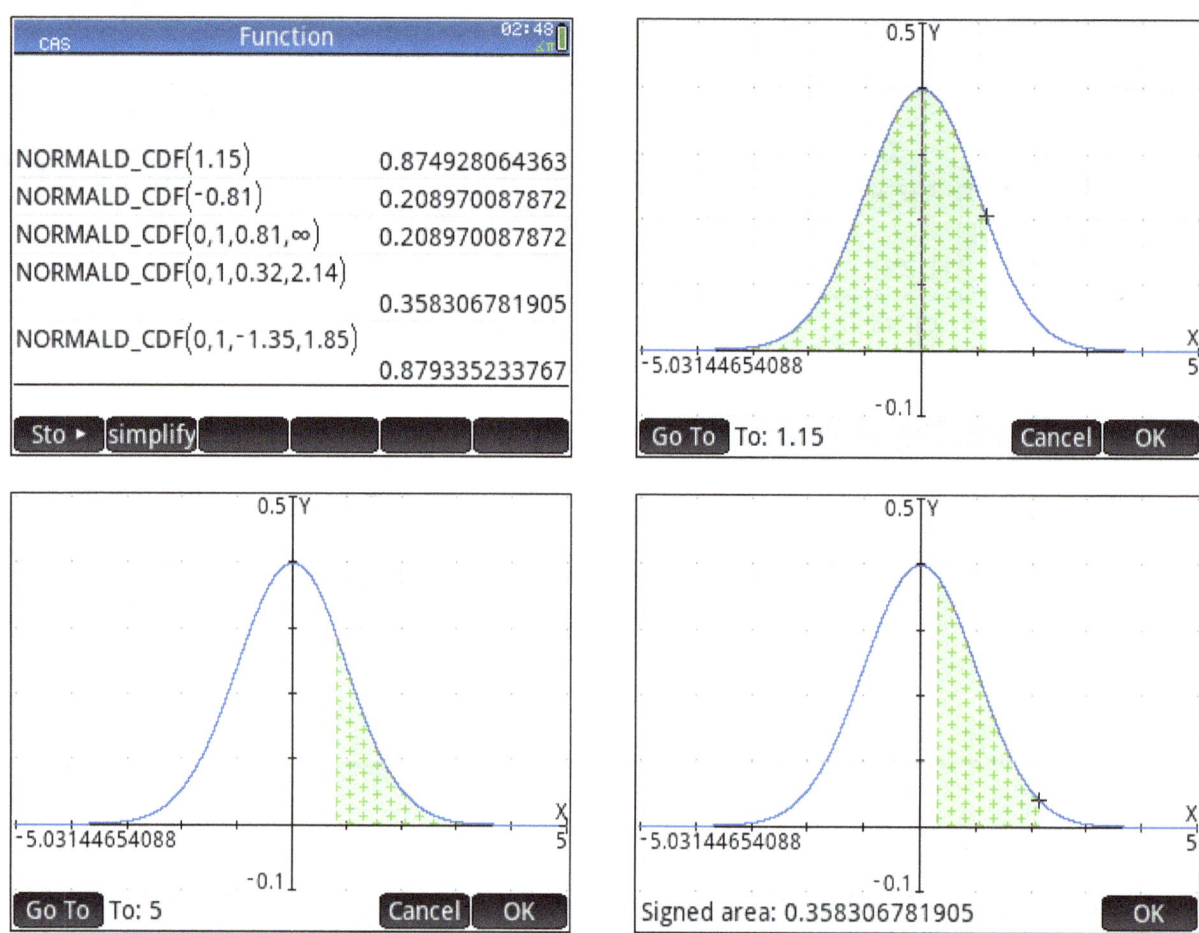

Z Scores

Sunday, August 18, 2019   11:25 PM

Z is the standard normal variable. Find the value of z that satisfies

$P(Z < z) = 0.9236$ $P(Z > z) = 0.9406$ $P(-z < Z < z) = 0.7498$

$z = 1.43$ $z = -1.56$ $z = 1.15$

$P(Z < z) = 0.9236$ $z = 1.43$

$P(Z > z) = P(Z < z)$

$P(Z < z) = 0.9406$ $z = 1.56$

$P(Z > z) = 0.9406$ $z = -1.56$

$P(-z < Z < z) = 2P(0 < Z < z)$
$P(0 < Z < z) = P(Z < z) - 0.5$
$P(-z < Z < z) = 2P(0 < Z < z) = 2[P(Z < z) - 0.5] = 0.7498$
$2\{P(Z < z) - 0.5] = 0.7498$
$P(Z < z) = 0.8749$ $z = 1.15$

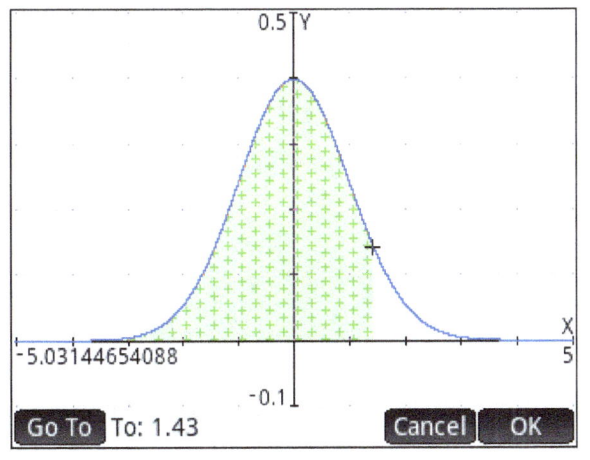
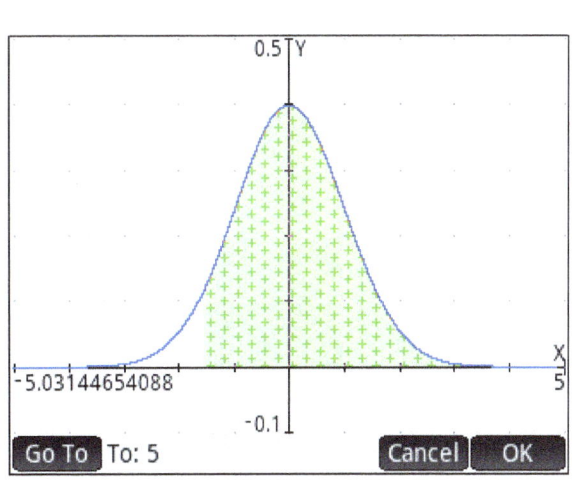

More Normal Distributions

Friday, August 23, 2019   3:33 AM

http://computerlearningservice.com/Academy/Finite-Math/Additional-Topics/More-Normal/more-normal.html

Other Means | Standard Deviations

Sunday, August 18, 2019   11:26 PM

Probability Density Function Normal Curve
The Probability Density Function for the normal curve,

$$f(x) = \frac{1}{\sigma\sqrt{2\pi}} e^{-(1/2)[(x-\mu)/\sigma]^2} \qquad Z = \frac{X - \mu}{\sigma}$$

Plot the area of $P(0.1 < X < 2.1)$ with mean, μ, of 1 and stsndard deviation, σ, of 2.

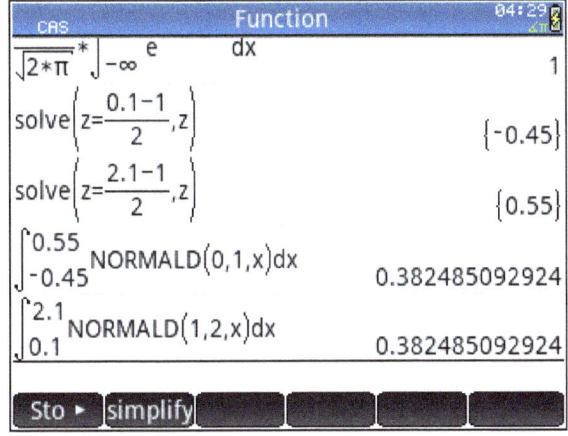

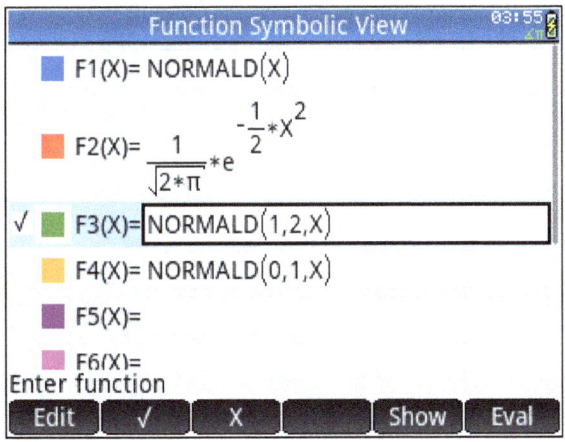

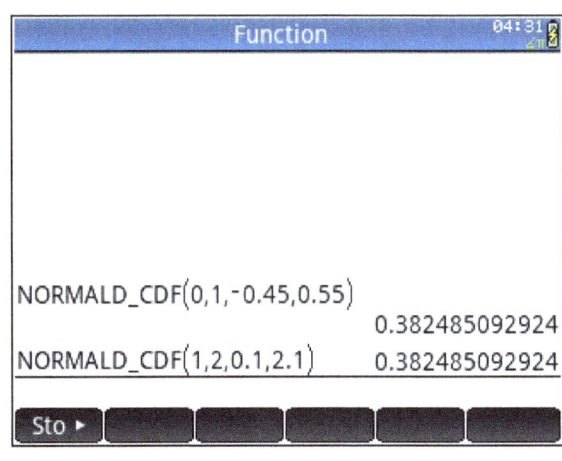

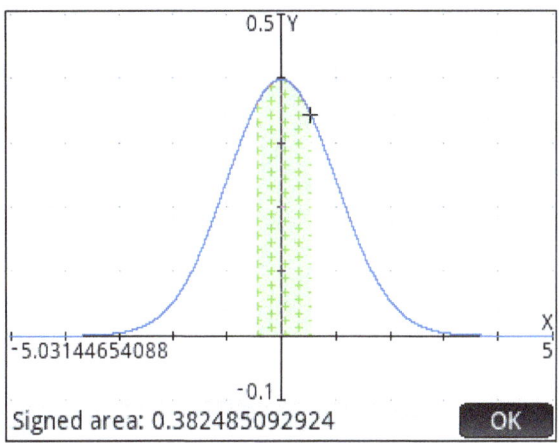

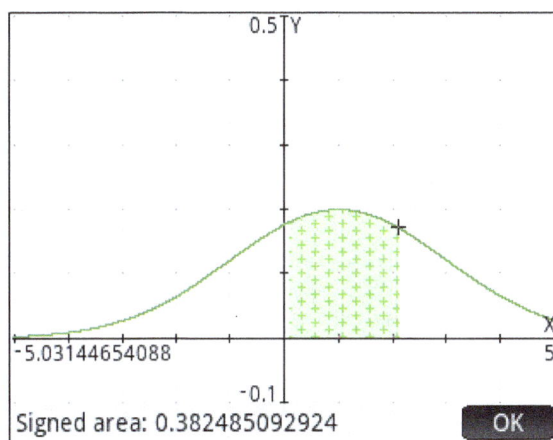

Additional Topics Probabilities Page 133

Subpage

Thursday, August 22, 2019   5:01 AM

X is a normal random variable with mean, μ, of 50 and standard deviation, σ, of 10. Find the value of x that satisfies

$P(X < x) = 0.9236$ $P(X > x) = 0.9406$
x = 64.30   x = 34.40

$P(X < x) = 0.9236$ x = 64.30

$P(X < x_1) = 0.9406$ $x_1 = 65.60$
$x = \mu - (x_1 - \mu) = 2\mu - x_1$ $x = 2 \cdot 50 - 65.50 = 34.40$

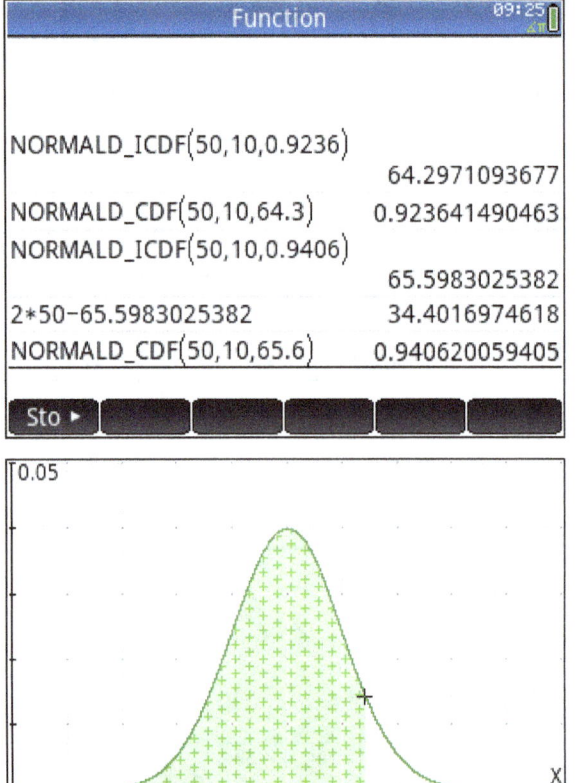

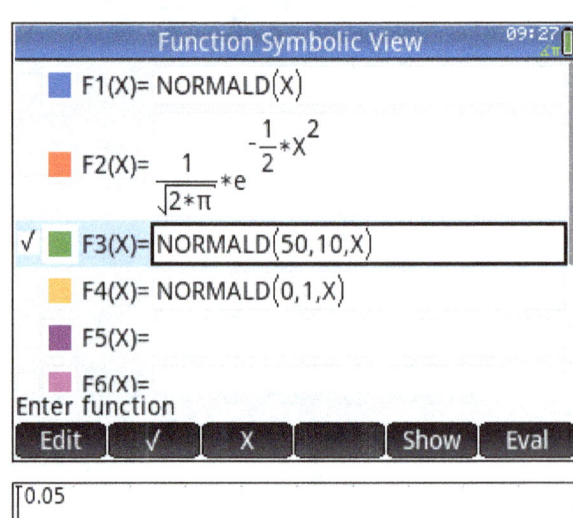

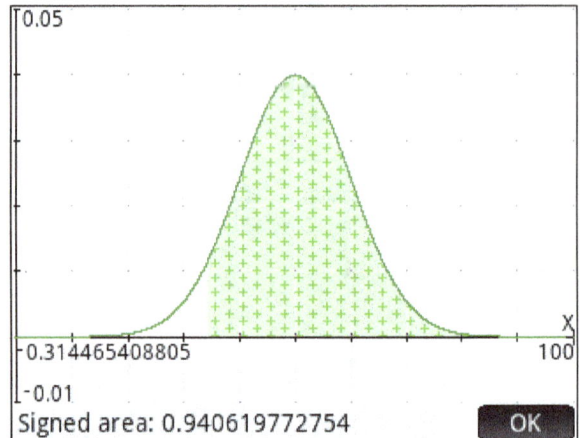

Approximate Binomial Distribution

Friday, August 23, 2019   3:36 AM

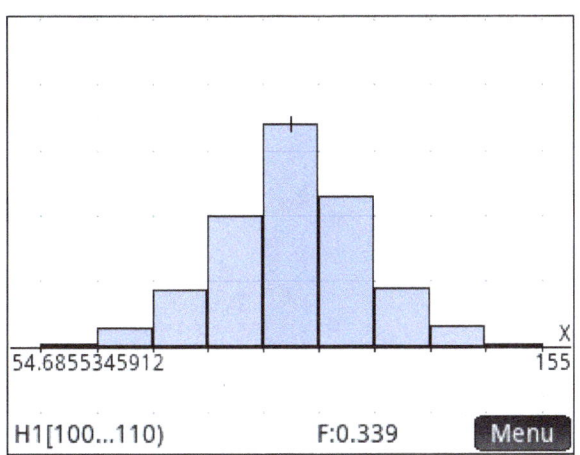
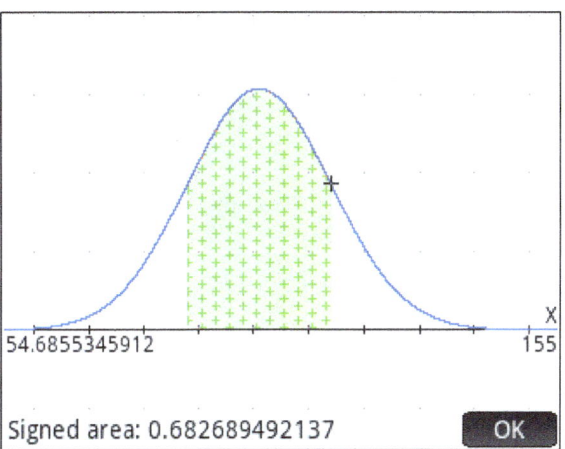

Stanford-Binet 1916 - Distribution of I Q's of 905 unselected children, 5-14 years of age.
How many children were with-in one standard deviation of the mean? $P(88 < X < 114)$
Answer: $905 \cdot 0.683 \approx 618$

How many children had I Q's above 114? $P(X > 114) = 1 - P(X < 114) \approx 0.159$
Answer: $905 \cdot 0.159 \approx 144$

www.ingramcontent.com/pod-product-compliance
Lightning Source LLC
Chambersburg PA
CBHW081357290426
44110CB00018B/2404